Edward Drummond Libbey

Also by Quentin R. Skrabec, Jr.
and from McFarland

Henry Clay Frick: The Life of the Perfect Capitalist (2010)

H.J. Heinz: A Biography (2009)

The Metallurgic Age: The Victorian Flowering of Invention and Industrial Science (2006)

Edward Drummond Libbey, American Glassmaker

QUENTIN R. SKRABEC, JR.

McFarland & Company, Inc., Publishers
Jefferson, North Carolina, and London

Library of Congress Cataloguing-in-Publication Data

Skrabec, Quentin R.
Edward Drummond Libbey, American glassmaker /
Quentin R. Skrabec, Jr.
p. cm.
Includes bibliographical references and index.

ISBN 978-0-7864-6335-0
softcover : 50# alkaline paper ∞

1. Libbey, Edward Drummond, 1854–1925. 2. Libbey Glass Manufacturing Co. 3. Glass manufacture — Ohio — History. 4. Glass manufacture — United States — History. 5. Toledo Museum of Art. I. Title.
TP853.O3S56 2011 748.2092 — dc23 2011030559

British Library cataloguing data are available

Front cover: (inset) Edward Drummond Libbey, 1920; (background) boys working the annealing furnace (both images courtesy University of Toledo, Ward M. Canaday Center, Libbey Glass Records)

Manufactured in the United States of America

McFarland & Company, Inc., Publishers
Box 611, Jefferson, North Carolina 28640
www.mcfarlandpub.com

To Our Lady of Fatima and
my granddaughter, Anna Shamrock

Acknowledgments

I would like to give special thanks to Julie McMasters of the Toledo Museum of Art and Kimberly Brownie, Ann Bowers, and Barbara Floyd of the Ward M. Canaday Center at the University of Toledo. These women are local treasures of knowledge and professional help as well as experts in the field. The staffs of both the University of Toledo archives and the Toledo Museum of Art archives were helpful and dedicated. Michael Ryan and Nicole Rivette were particularly helpful at the museum. The many volunteers and staff at the Toledo Museum of Art provided encouragement and education. The museum remains today a tribute to Libbey and his wife, Florence Scott Libbey. Thanks to the outstanding staff at the Toledo-Lucas County Public Library, in particular Ann Hurley, Greg Miller, James Marshall, and David Lora. Additional important help came from Lee McLaird of Bowling Green State University's Rare Books and Special Collections. The staff at the Rossford Public Library were extremely helpful, particularly Anne Bushel, Charles Butcher, and Lawrence Bridenbaugh. Another important resource was the Hancock Historical Society and its director, Paulette Weiser. The staffs at local libraries in Tiffin and Findlay, and at the Fostoria and Hancock county historical societies were also extremely helpful. The Northwest Ohio area is blessed with many knowledgeable glass clubs in Toledo, Tiffin, and Fostoria, which were also sources of information.

Another important facet of the research was done at the Rutherford B. Hayes Presidential Center archives; I would like to thank the Center's Nan Card, chief archivist, and Merv Nall. Other archival resources were the Henry Ford archives and the Benson Center archives, which holds the corporate records of Boston and Sandwich Glass. All of these resources are old friends and have served me well over the years.

Table of Contents

Preface

For more than twenty years, I have been researching and writing about a number of key capitalists, politicians, industrialists, managers, and educators who contributed to America's industrial revolution. The inductees of this special pantheon are representative of the golden era of industrial America. Among its members are H.J. Heinz, George Westinghouse, President William McKinley, Henry Clay Frick, William McGuffey, Michael Owens, and Andrew Carnegie's "Boys of Braddock," but it still lacked an industrialist that pioneered the model of today's chief executive officer (CEO). Edward Drummond Libbey, more than anyone meets that model. Libbey not only founded a number of major American companies such as Libbey Glass, Libbey-Owens, Owens Bottle, and Toledo Glass, but also integrated his Toledo Museum of Art into the community. At his California ranch, he built a new cultural small town known today as Ojai, California. Libbey brought art, industry, and community together in unique ways that offer insights for today.

Amazingly, Libbey has avoided the pen of biographers, and thus the book offered a new kind of challenge for me personally. There was good reason for Libbey's lack of notice by biographers. He left few personal notes, unlike many industrialists of the era. What little personal writing, letters, and notes he did write appear to have been destroyed on his death by his own request. The items that did survive are his early glass formula notebook, a brief "travel log" of his early years (fewer than ten small pages), and a handful of letters. These are located today at the Toledo Museum of Art. Many of the company records of Libbey Glass and Libbey-Owens have only recently been fully cataloged for research by the University of Toledo. I was able to take advantage of some remaining "community memory" of Libbey, his early companies, and the legacy of his museum.

This book offers a unique and first detailed look at one of America's greatest industrialists. Until Libbey and his partner automated glassmaking with their bottle machine, the craft had not changed in three thousand years. Libbey's companies would eventually automate all types of glassmaking, which made glass the most common packaging material and brought light to homes and buildings with massive windowpanes. Libbey's companies remain an important part of the American industrial landscape today. Just as important is that Libbey glass pieces are the most prized items for glass collectors. The Libbey name remains synonymous with American glass to this day. His famous punch bowls and art pieces are prominent in museums across America.

The book also offers a look at one of America's premiere art collectors and his quest to bring his collections to the American public. Libbey went far beyond collectors such as J. P. Morgan and philanthropists such as Henry Clay Frick in demanding a living tie to the community of his museum. His community museum building model rivals his corporate business models.

Certainly, Libbey's business career would have earned him a seat in any pantheon of American industrialists, but he brings a very unique vision. Libbey represents a tie between industry, community, and art not seen in other industrialists. Edward D. Libbey brought his executive skills to many endeavors. He was a "gentleman owner" who defined the mission and vision of his company. His management was always strategic, leaving the tactics to his team and leaders such as Michael Owens. In many ways he was bred to be an owner, not by an aristocratic education, but by experience. Libbey best represented the ideal relationship between community and company. His philanthropic efforts were not aimed so much at giving to the community as at involving the community. The Toledo Museum of Art today remains Libbey's legacy of bringing out the best in an industrial community. He was an unabashed capitalist, but one who looked to improve the world with his money.

Libbey shares many of the characteristics of his fellow industrialists. He was a capitalist in every measure. He accumulated wealth, but he proved a fair employer. He shared the vision of President William McKinley's alliance of labor and management. He whole-heartedly supported the protective tariffs of McKinley's "American System." Libbey had been part of McKinley's many trips to the Toledo Club to discuss industrial policy. He showed the responsibility expected of industrialists to invest profits from the tariffs into factories, research, and American jobs. He believed in industrial peace at all levels, although his methods of monopolistic control might be questioned. He expertly managed the transition of the glass industry from a handcraft to industrial automation. He held true to the *McGuffey Reader* principle that capitalism must give back. Giving was a responsibility, but the nature of that giving was an individual decision. The Toledo Museum of Art is the manifestation of his sense of obligation.

Introduction

The Toledo Museum of Art affected me personally in the late 1960s. My family moved from Pittsburgh to Adrian, Michigan, as my father took a job as vice president of Green Manufacturing. I was just starting high school and didn't really care for the move. My father took me to three places he knew I would enjoy in the area—Greenfield Village, the Toledo Museum, and the University of Michigan. All three would put me on the path to write about capitalism, but the Toledo Museum of Art was part of my early career choice. For years I had visited the Carnegie Institute and Museum on an almost weekly basis. But the Toledo Museum of Art seemed much different. I had little interest in art but found the sense of art, design, and materials different. It changed my focus from pure science to engineering and materials. Four years later, after many visits to the museum, I entered the University of Michigan to pursue a career in metallurgical and materials engineering. I often return to the museum today and find that original fascination in walking the galleries. I love the broad focus that was far different from my first trip to an art museum in Pittsburgh.

Years later I returned to Toledo to get my Ph.D. in manufacturing, and the museum once again supplied motivation. That may seem strange, but that was Edward Drummond Libbey's vision, and one I fully understand. It was this unique vision of Libbey's that made writing this biography such a pleasure. It is what makes the Toledo Museum different from most art museums. That spirit exists today and can be sensed on a visit to the museum. It confirms that an art museum belongs in an industrial town as much as in a metropolitan city. The glass collection provides inspiration to the manufacturer as well as to the artist. The glass industry is one of the few where we see this fusion of art and industry. It is this spirit that has launched several revolutions in industrial design and art movements. It has brought engineering classes to the museum because it educates on design. The museum today remains true to Libbey's vision.

Libbey's life offers much more than a contribution to art and design; it is a textbook for marketing and development managers. Libbey developed a universal, organizational approach to success. His encouragement of creative freedom served as a model for many industrial companies and also served well the museum and urban planning of Ojai, California. He borrowed much from Andrew Carnegie and John D. Rockefeller in his organizational design, but he pioneered the role of CEO. He developed a balance of hands on/hands off management from the top. He pushed his organization without making it dependent on him for day-to-day operations. He fostered creativity in his managers and was able to find room for the problems very creative people bring to organizations.

In Mike Owens, Libbey found and nurtured creativity while allowing the eccentric behavior that often comes with it. The partnership of Owens and Libbey often tortured traditional managers but proved to be one of America's most profitable. The only other partnership that came close was that of Nicola Tesla and George Westinghouse. Tesla's eccentric lifestyle forced him out of Edison General Electric into Westinghouse Electric. Owens similarly would have been incompatible with most business organizations. Libbey, like Westinghouse, understood that creativity requires more freedom.

Libbey is a role model for capitalists. He gave back to the community. Many might argue that he should have given to the poor instead of building an art museum. It is an old argument, but the world needs museums as much as it needs soup lines. More importantly, how one gives back to the community is an individual matter in our society. Libbey was blessed that his passion for art, love of glassmaking, community vision, and hope in education came together in a single project. His focus was on children and education, which benefited society. Libbey approached philanthropy differently than most other capitalists. He wanted true community involvement in his giving. He didn't just give Toledo an art museum, he made Toledans part of the building. He made the museum a community center for education. Art and industrial art education were part of the museum's mission. Libbey found encouragement, not just in the number of visitors to the art galleries, but in the number of students enrolled in training at the museum. He even promoted visits by workers and industrial designers to infuse art and manufacturing into the curriculum.

The museum today remains a spiritual legacy to Edward Libbey. Walking through, you can sense his spirit and direction. The glass collection is a tribute to the glass industry, and a living example of the fusion of industry and art. The diversity of the collection is striking in itself and reflects the central role of art that Libbey envisioned. The museum is still inviting, as Libbey had envisioned. It is in every aspect an art museum for the average person.

On a personal level, Libbey was one of those rare individuals who achieve self-actualization. He found a type of integration in his life, while most of us have boxes or segments that divide our careers, personal lives, and passions. It was not easy. Libbey's life shows a struggle for that integration of goals, often reflected in his search for the mission of his museum. In business, he found a rare enjoyment. He tried to retire but was always pulled back. Still he was able to find a balance of work and relaxation. His friendships and relationships were deeply private. He kept few written records and what he did keep, he requested his wife and lawyers to destroy on his death.

Chapter 1

A Crisis in Frogtown

The hot, humid summer of 1888 was typical of the swampy city that had come to be called Frogtown. Many parts of today's downtown Toledo were described as a marshy jungle in the summer months. Water was everywhere and it was hard to distinguish drainage canals from streets at times. The canals often doubled as a sewage system. This created a biblical plague of frogs every summer. Often as he walked, Edward Drummond Libbey would sort through the mail he had just picked up from the Toledo city post office. And often he was disappointed in the number of incoming customer checks. It seemed as though he and his father before him had been in this position forever. Edward was broke, but on top of it he suffered from homesickness. The stress was wearing him down physically.

He carefully followed the wooden planks through the mud. The city was still suffering from the spring flooding of the Maumee River and mid-summer rains. His heavy coat did nothing for the heat but kept the swarms of mosquitoes away. He was returning to his new living quarters at the Madison Hotel to meet his sister. His glasshouse was only six months into operation, and Libbey wasn't sure if he could meet the weekly payroll. His personal funds were exhausted, and the money his sister Sarah had lent him was now in jeopardy. He also had money from friends as well as hundreds of small investors in his new hometown of Toledo. Libbey had always tried to use family, personal, and small stock subscriptions instead of heavy involvement by banks.

Nearly half the skilled glassmakers who had followed him from Boston had returned to Boston. The Irish workers who had stayed seemed more interested in drowning their homesickness and hatred of the unrelenting humidity and heat, endless plagues of frogs and mosquitoes, and frustrating plant operations in the barrels of the local Findlay Beer. The boy laborers were undisciplined and crammed in overcrowded boardinghouses.

Nationally, business was expanding, as was the demand for glass, but Libbey was not realizing the benefits of this prosperity. The quality of Libbey Glass, which had been the finest glass when manufactured in East Cambridge, Massachusetts, could

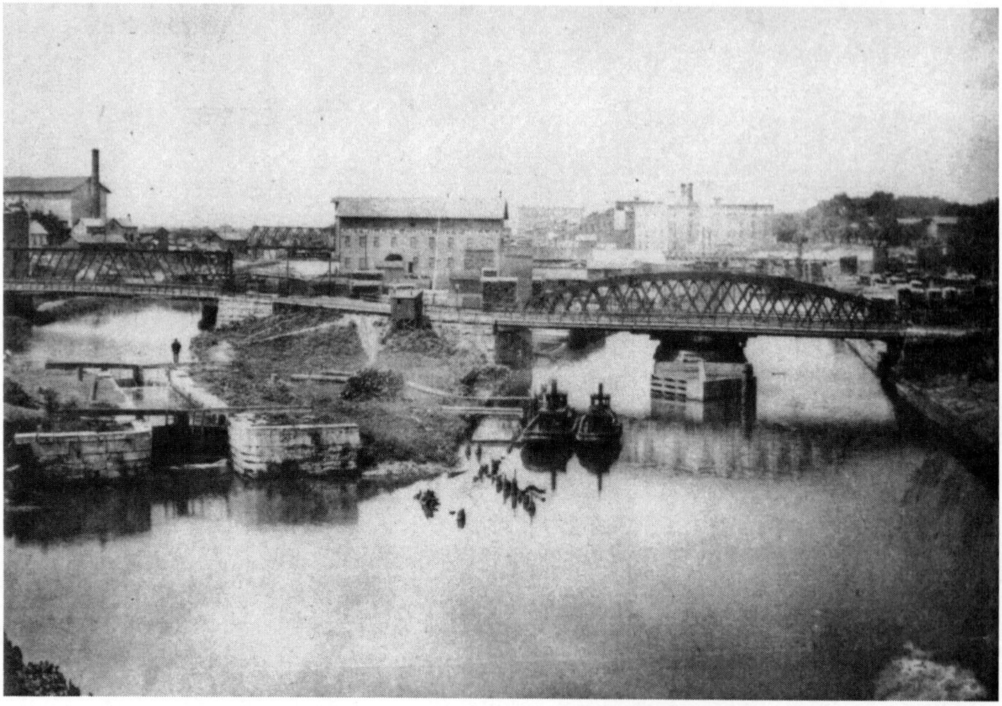

Toledo (Frogtown) at the time of Libbey's arrival (courtesy of University of Toledo, Ward M. Canaday Center, Libbey Glass Records).

not be achieved at the new Toledo glasshouse. Libbey's famous flint glass formulas, which had been the envy of his competitors, did not seem to work at the Toledo glasshouse. The new multi-fuel furnaces were having major problems. The workforce was often too hung over to get the furnaces ready for the week's production. Crews were short skilled positions. The Toledo area offered no replacements for the lost glassmakers, only unskilled women and children. Some workers went back and forth from the local saloons to the glasshouses. Drunken workers often started fistfights with their foremen.

The mail included more bad news from his Boston lawyers. He was still trying to settle stockholder lawsuits from East Cambridge which claimed mismanagement by Libbey caused the failure of New England Glass. Libbey's optimism of February 8, when a *Toledo Blade* headline hailed, "Glass Is King," had evolved into despair. One benefit of failure was that he could return to his beloved Boston.

In 1882 he had taken over the company on the death of his father, William, who had struggled for thirteen years to save the company. The company had roots going back to 1818, but the economics had shifted. The rich hardwood forests of New England, which had been the primary fuel of the company, had been lumbered out. Cheaper fuel sources in Pennsylvania, West Virginia, Ohio, and Indiana had created more efficient competition. While labor represented around 55 percent of controllable costs, fuel was next, at about 33 percent industry-wide.[1] In New England the numbers were nearly reversed, making fuel the main cost.

Most of the larger companies and successful glassmakers had left New England in the 1870s, but the Libbeys held on. The Libbey family could actually trace its lineage back to the *Mayflower*. This was their country, and 240-foot smokestack of New England Glass was a landmark. Its lack of smoke was a symbol of New England's manufacturing decline. The Libbeys were part of the Boston gentry and were appalled at the thought of moving to the American frontier. Boston rivaled New York as a home for the country's elite; but it was also a solid, blue-collar town. Baseball had become the city's pastime, and glassworkers formed teams and watched the city's professional team. But the industry was moving west; by 1870 Pittsburgh had become the glass capital of the nation.

Many of William Libbey's friends had moved on to Pittsburgh and West Virginia, but he hesitated. He had revived the fortunes of the company in 1883 with the introduction of his famous cut colored glasses of Amberina and Peach, making the company profitable for the first time since 1873. William and his glass masters had developed these new colors just prior to William's death. Still, labor and fuel costs overwhelmed his innovations and the marketing success of these new glasses.

It was Edward Libbey who took samples of the new amber glass to retailers and successfully marketed it. He rode the rails all over the East Coast and set up sales outlets for the new colored cut glass. In 1884, Edward sold the company's entire stock to Tiffany's of New York. The beautiful reddish glass known as Amberina resulted from a rediscovery of using gold to color glass. Its success turned the company around, and New England Glass reported a profit in 1884 for the first time since 1873. Libbey, sensing a new competitive weapon, introduced several colors and heat sensitive glasses, such as Agate, Peach, Peachblow, Burmese, and Pomona. Metal oxides and special heating produced a mix of colors. Uranium oxide was used to produce yellow, and copper oxide, green. Libbey combined these new colors with unusual art designs. Sales boomed. Orders were late or unfilled.

From 1883 to 1886, Edward Libbey used the cash from patent and product sales to reduce debt. The first piece of Amberina, a fluted bowl, can still be seen at the Toledo Museum of Art, as well as the first Pomona cream pitcher. Libbey was able to sell the rights to making heat sensitive glasses such as Amberina to other big producers like Hobbs-Brockunier in Wheeling. Peachblow was an extremely beautiful glass made with cobalt and copper oxides. The Hobbs-Brockunier product was sold as Wheeling Peachblow and Coral. Libbey was not the first to patent glasses and designs, but he defended them better than most. Libbey also won lawsuits when others tried to copy his patents, such as when Mt. Washington Glass produced Peach Skin. Even these proprietary formulas, however, failed to produce, in the new Toledo glassworks, the beautiful glass that Libbey had sold America on. But like his father, Edward had proven to be a master marketer.

The financial success of these products allowed Libbey time for a vacation. From May to September of 1887, Edward Libbey made a grand tour of Europe, which he believed was necessary for a glassmaker. Libbey managed to sell patent rights for Amberina as well as tour some of Europe's greatest glasshouses such as those at Murano. Unfortunately, the success in Europe resulted in cheap copies being shipped to the United States.

The trip offered Libbey is first chance to study European art, which would become a lifelong passion. He became a collector as well as a manufacturer. European glassmakers had changed their process little over the centuries. Libbey strongly believed he was a manufacturer of art rather than an industrialist, but things would soon be changing, as the cut-glass market continued to decline.

Libbey wondered whether he was not cut out for the operations end of the business. Cost pressures from fuel and labor continued to offset sales, and the company was in the red again by 1886. Glass firms were mushrooming throughout Western Pennsylvania and the panhandle of West Virginia because of fuel availability. Western Pennsylvania was blessed with both coal and natural gas. By the early 1850s, good hardwood had been lumbered out and a new fuel — coal — offered better efficiency. Some New England firms successfully switched to softer woods such as pine, but the consumption of wood by these glass factories was enormous. A factory could easily consume ten thousand cords of wood in a year. In good times, a glass factory would consume a football field of forest in a week! A century earlier, England had banned the use of hardwood in glasshouses because the Navy needed it for ships. Coal was abundant in Pennsylvania and West Virginia, and that abundance fired the emerging glass industry of the area. Glass historian E. William Fairfield summarized the situation:

> The proximity of big coal supplies in the West immediately gave New England's Western competitors a powerful economic advantage. By 1860, the once plentiful wood supply in New England had dwindled ... hundreds of acres of timberland that had been leased by the New England Glass and Boston & Sandwich had been stripped for fuel. Even had there been hundreds of additional acres available, wood as fuel was no match for the efficiency of coal. The potash (potassium carbonate) once secured from wood ash had been replaced in the glass formulas by chemicals, which proved superior. It took great quantities of wood, oil additives, and powerful drafts to make the wood fires intense enough to melt and work glass. New England Glass Company's factory's chimney ... towered 235 feet in the sky.[2]

Libbey had come to Toledo determined never to be caught in a regional fuel crisis again. Toledo offered supplies of coal, oil, and natural gas. Libbey's new glass furnaces were built to run on any of the three fuels. While abundant, coal changed the nature of the glass industry, and Toledo's railroad connections made coal in Toledo as cheap as in Pittsburgh. Two types of coal were available, a hard coal known as anthracite and a soft coal known as bituminous. Anthracite could be used as a fuel directly while bituminous needed to be converted to coke, by burning it in an enclosed oven to remove impurities and produce a dense fuel. Coking was labor intensive, and, like coal mining, became an employer of cheap immigrant labor.

Even though coke in the glass furnaces produced a hotter and more consistent heat than wood, there was a price. Coke produced a residue known as clinkers, which had to be removed. The bigger drawback was that coke was dirty. The finished glass product required washing, cleaning, and packaging. A glass manager in 1886 described it this way:

> When the ware is pressed and placed in the leers, which are heated with coke, it comes out very dirty, keeping a number of hands busy placing the glass in carrying boxes; then to the wash room, where it is washed one piece at a time, then repacked in the boxes and sent to the packing room. In double handling of ware a great amount is broken.[3]

Libbey's main business was the fine cut and art glass market, and dirt could not be tolerated. Coal in general was a cheap but dirty fuel. In art and clear glass factories, pots of molten glass needed covers to prevent ash and cinders from entering the glass.

Coal was driving a glass industry exodus from New England. Pittsburgh was the first to benefit. Its glass industry was dominant from 1850 to 1880 because of the nearby coalfields. Coal heating could substantially increase productivity by producing more "metal" per hour with larger glass melting pots. Pittsburgh was naturally rich in coal reserves. It was often shoveled directly from open seams in the surrounding hills. The Pittsburgh industry was proud but arrogant, driving some of New England's glass masters to the West Virginia panhandle. The success of the Western Pennsylvania coal companies and the expansion of Toledo as a major railroad center had made coal in Toledo as cheap as in Pittsburgh. Similarly, Rockefeller's Cleveland refineries and the railroads had made oil cheap in Toledo as well. Oil was flowing by tank car from the refineries in Cleveland. In the 1880s, Toledo's energy costs across the board were the lowest in the world. Because of his painful experience in wood, Libbey was one of the first businessmen to fully appreciate Toledo's potential.

By 1886, Hobbs, Brockunier & Co. had become one of the first companies to successfully use natural gas as a fuel. Actually, Rochester Tumbler Works near Pittsburgh lays claim to being first to use gas in 1875. Gas offered some amazing advantages over coal. It could supply higher heat than coal at about a third less cost. This new fuel further strengthened their cost advantage over such competitors in New England and Pittsburgh. The use of gas allowed for more "metal" per hour by increasing the size and number of pots. An average coal- or wood-fired glasshouse might have six pots of 700 pounds each. With gas, the average glasshouse had ten pots of 2,500 pounds each.

Libbey hoped to capitalize on this advantage. What he hadn't counted on was the effect on his chemical formulas. The higher heat of gas changed the subtle chemical reactions needed to perfect the color. The wood-fired formulas had been developed over centuries, and now adjustments were needed in a matter of weeks. Some of his best glassmakers had returned to Boston as these critical adjustments were under way. While the formulas never left Libbey's possession, which was typical in the industry of secrets passed down from the family owners, Libbey himself was not a master glassmaker or gaffer, even though his father had been. The formulas of other companies had been adapted, but Libbey Glass lacked operations experience. Gas, however, was driving down the costs of glassmaking and to be competitive, a solution had to be found.

Toledo had pooled its resources to build a natural gas pipeline to attract Libbey. The gas field was centered in the town of Findlay. Had Findlay's gas supplies been extensive, it might have taken the title glass capital from Toledo. An amateur Findlay geologist, Dr. Charles Osterlen, had observed oil slicks and "marsh lights" for years. These were similar to the sightings around early Pennsylvania wells; the first successful oil well was in Titusville, Pennsylvania, in 1859. In 1884, Dr. Osterlen's drillings hit gas under Findlay in the Trenton limestone formation at such pressure that the supply seemed endless. The Trenton limestone formation ran under most of Northwest Ohio and parts of Indiana, bringing many growth-oriented towns into the competition.

The real gas boom started, however, when Findlay butcher Louis Karg struck gas on the south side of the Blanchard River. This strike roared into being, lighting the whole of Northwest Ohio. The Karg well was of enormous commercial value with a flow of 20 million cubic feet daily. Optimism broke out all over Northwest Ohio with offers of free gas if companies would bring their factories. The size of the Trenton limestone formation suggested a huge reservoir. Local entrepreneurs found support in some more radical geological theories. On August 11, 1885, *The Toledo Blade* quoted Professor J.P. Lesley of the Pennsylvania Geological Survey as saying that gas formations were self-generating. Furthermore, he suggested that "it would be impossible to sink a well any place within this basin without finding gas, if the well were drilled to the proper depth."

Test drilling accelerated throughout the area, with oil being found in Lima. Eastern capitalists rushed to the area with cash and plans. Boomtowns such as Fremont, Fostoria, North Baltimore, and Bowling Green sprang up. Glass companies, in particular, were attracted to the cheap source of fuel. Glass tended to be the most mobile industry. Unlike steel, oil and others, glasshouses required little permanent machinery and could be moved by rail. The framing of the plants was wood, which could be found locally. Between 1870 and 1910, it was common for whole glass plants to be moved to areas of cheap fuel. Libbey had filled 25 train cars with equipment as he brought the East Cambridge factory to Toledo.

While hundreds of companies rushed to the heart of natural gas production in

Findlay, Ohio, lured by offers of free gas, Libbey looked for a location that could supply alternative fuels as well. John D. Biggers suggests that Findlay was high on Libbey's list, but the Toledo Businessmen's Association won him over.[4] Libbey had been offered free gas in Findlay, but he was determined never again to be hostage to one type of fuel. Toledo planned to pipe the natural gas from fifty miles away in Findlay, but Libbey gained fuel flexibility that would be needed in a few short years.

In 1888, Toledo was a railroad-lake-canal transportation center that rivaled Chicago. That would allow Libbey to buy coal at Pittsburgh prices if needed. Toledo's proximity and rail connection to the booming oil refineries of Cleveland could supply oil at the best prices. Toledo therefore had access to natural gas, coal, and oil. Between 1885 and 1890, more than two hundred firms built glassworks in the Findlay/Bowling Green area, but Libbey boldly (and in hindsight correctly) chose Toledo. His decision came after extensive study as he tried to justify leaving his beloved Boston. Toledo had recently discovered a vein of flint sand which was similar in the quality to New England sand. Fuel and raw materials were fundamental, but success depended on the glassworkers who put it all together; and even in Toledo, Libbey had to deal with an emerging union. The union, as much as fuel costs, had contributed to the decline of New England Glass. Libbey brought his labor problems with him to Toledo.

Glassworkers were the aristocrats of America's skilled workers. The predecessor to Libbey Glass, New England Glass, included a secret organization of skilled glassblowers. Most of the flint workers were outside this group. The structured system consisted of a "gang" headed by the blower or master craftsman, followed by the assistant blower, gatherer, assistant gatherer, and servitor. A glasshouse might have 8 to 14 gangs working on various shifts and furnaces. In addition, the gang was supported by a group of boys (usually 4 to 8) that handled the glass in various stages of production. The first effort to unionize flint glass workers was in Pittsburgh in 1858. The organization was known as the Glass Blowers Benevolent Society and focused on the blowers.

The Benevolent Society continued as a secret organization without much success. As with most blower unions, it lacked the ability to strike successfully against a company that would use gatherers or non-union workers to fill in. It wasn't until 1878 that the Knights of Labor brought a number of eastern unions together in the United Flint Glass Workers. This union now represented the whole gang (only the handling boys, making up the biggest fraction of employees, were left out) and was the first strong union to address the art glassmakers such as the Hobbs-Brockunier Company and Libbey Glass.

The Flint Glass Workers in New England reorganized in 1881 as an American Federation of Labor (AFL) union. The Flint Glass Workers unionized the art and cut glasses factories such as Hobbs-Brockunier, Libbey Glass, and Corning Glass. The Flints tended to be more focused on the apprentice system and schedules than wage demands, but they wanted geographical parity in wages. This focus was natural since the Flint Glass industry best represented the artisan tradition of Venetian glassmakers. The Flints modeled their union on Venetian glassmaking principles and the ancient organization of the gang. There was a feeling of superiority and pride in their segment of glassmaking. But these Flint Glass Workers were not Venetians, but hard-drinking Irishmen.

Edward Libbey had a great respect for the traditions of the union, but to stay in New England required wage concessions, not parity with western glass companies. Libbey had continued to look for ways to remain in New England right up to the minute he stepped on the train to Toledo. He was facing a new strike at the East Cambridge plant in the months prior to leaving.

Even as the glassworkers were highly paid, the union had a pay issue. The fight centered on a discrepancy between the pay scales of the old Eastern glassmaking districts of New England and the more profitable mid–Atlantic and Western districts. The cheaper fuels in Pittsburgh, West Virginia and the Midwest allowed higher wages in the West. Shops in Ohio, West Virginia, and Pittsburgh were paying $80 to $100 a month for skilled workers, which was 25 percent more than the New England shops were paying.

The American Flint Glass Workers, who had been part of those wage increases in the Midwest, were urging East Cambridge workers to stand fast. The union did not or would not consider the difference in manufacturing costs between the two areas. In addition, the Midwestern branch of the union dominated. It wanted solidarity even at the cost of plant closings in the East. In fact, one of their leaders, young Michael Owens, would play a key role in the ultimate future of Libbey Glass.

Libbey felt uncomfortable discussing company financial issues but he had no choice. Libbey personally negotiated with the workers, showing the serious financial position of the company. In the end, the workers struck under pressure from the union national headquarters, dealing the final blow. This was all he needed to make the decision. Libbey actually understood their position, and in the end he would meet the American Flints' demands at his new Toledo plant. This in hindsight looked like a major mistake but Libbey understood that they made the glass, not him. In the end, Libbey not only paid for the workers to move to Toledo but paid to move their families as well. Now he was being second-guessed by some of his employees.

Sarah would become his best advisor in his personal and company finances. She was a source of much needed strength. She pointed out how his father, decades before, had risen above the tribulations of the changing glass market to save the company. Libbey was borrowing heavily in the early months on his personal accounts, and now those were dried up. Sarah believed his main problem to be a simple case of homesickness; Toledo was no match for metropolitan Boston.

Libbey wasn't alone in his homesickness. The predominantly Boston Irish workforce was leaving as well. Libbey had invested in baseball fields and equipment to get his workers' minds off Boston. The workers loved baseball, but Toledo was far from the baseball centers on the East Coast. Toledo's minor league team, the Black Pirates, was no substitute for their beloved major league Boston Beaneaters (Toledo's Black Pirates had just been taken over by George Ketcham and his brother, Valentine Jr., who opened Sperango Park at Cherry and Franklin Streets). The workers missed the drinking and bars of Boston. The humid, muddy fields of Toledo were unlike the fields of Boston. As workers returned to Boston, Libbey was forced to use unskilled women and boys (if he could find them). Toledo lacked an industrial workforce, and Libbey started advertising in other cities for workers with large families of young boys. Short of cash and skilled labor, Libbey Glass seemed doomed. And Libbey wasn't alone; the Businessmen Association and local bankers that had supplied capital and land were deeply concerned.

Sarah continued to push Edward into his strength — a plan to save the company. Sarah had never wavered in her belief that the move to Toledo was the right one. She had seen her father struggle in New England and remembered how Edward had often advised him to move west. The Toledo banks could provide more capital and some family assets could be used. What Libbey lacked as an operations manager, he made up for in business. He had marketing, administration, and financial managers but lacked works managers to resolve the tough problems, which were all in the operations end.

Finance, investment, marketing, and recruiting were Libbey's personal cornerstones, but he had limited operations experience. He planned a number of recruiting trips to Pittsburgh and Wheeling to find skilled replacements for the men he had lost, but he still would need an operations manager. Libbey's right-hand man, Solon O. Richardson, was a brilliant administrator, a skilled process control man, and a follower of the famous industrial engineer Fredrick Taylor, but the factory had gone past the break point to total disorganization. Administration and process operating standards could come only after order could be established.

Joseph Locke, the plant superintendent, was a famous designer who had developed the Pomona, Peachblow, Amberina and Agata art lines. He also seemed to ignore the problems. Like Libbey, Locke represented the aristocratic gentleman glassmaker of a previous era.

Locke was one of the greatest glass artists of the time, skilled in the chemistry and artwork. As plant superintendent, Locke had full operational authority but preferred to do artistic work. Locke's authority covered both the hot and cold end of the business, but Locke let John Hopkins run the glasshouse with little interference. Locke was a great artist but no manager. The August 18, 1888, edition of *The Toledo Blade,* on the arrival of Libbey Glass in Toledo, said of Locke: "He is genial, popular, and obliging, and is liked by many of the 500 men employed by the company. He is a talented man in other lines, being an artist and musician of considerable reputation in Boston." The dual job of artist and manager presented a problem in the highly competitive glass industry. Locke, like Libbey, viewed the glass industry as a trade of gentlemen. Managers rose through the guild ranks based on their skill at working glass. But managing men required a completely different set of skills. In Libbey's organization, the company's best artists were also managers yet few were capable.

The floor manager or plant manager reporting to Locke was John "Gaffer" Hopkins. Hopkins was a fifty-year-old ex-glassblower and artist. Hopkins and Locke had managed well enough in the tranquility and prosperity of New England, but the economic challenges of the new location were overwhelming. John Locke, Jr., was superintendent of the etching department. The cutting department foreman, Willie Anderson, was also an award-winning artist. New England production methods were part of the problem, but the work ethic was also to blame. Neither Libbey nor Locke nor Hopkins could handle the labor problem.

Unfortunately, Toledo's natural advantages were offset by technical problems with the furnaces and the labor issues. By the fall of 1888, after only three months of operation, Libbey had lost 35 skilled glassmakers, which had crippled production of his art glass. What boys he could find were undisciplined and often ran away for a couple of days to go to the circus or on holidays. The men who stayed drank too much. The men argued more than they problem solved. Worse, the furnaces were not working well with Libbey's glass formulas. The new gas heat could not be controlled. Poor glass quality caused inspection rejections to mount to the point that the factory was producing only a few pieces each day, and orders were delayed.

With cancelled orders and late deliveries, income dropped off quickly. Libbey's credit line was exhausted and he needed to find more money. He didn't like banks and preferred private investors. Libbey had no ties with New York banks, which controlled most of America's capital. Furthermore, his father had warned him of the dangers of too much bank credit. Still, Libbey had few options with all of his personal money already in the firm. His sister Sarah hoped to raise more money from family and friends.

Sarah Libbey had started her brother on a plan to turn things around, but his infatuation with a local girl, Florence Scott, made the difference. Florence was a striking girl with a pedigree going back to New Haven colony in Connecticut. The Scotts were one of the founding families of Toledo. The Scott family had started *The Toledo Blade*, the free library, the park system, and Toledo's famous Manual Training School, and donated the land that would become the Toledo University of Arts and Trades. Florence had been educated in Europe and was knowledgeable in the fine arts. She was the type of girl that Libbey could have expected to find in Boston. His courtship of Florence helped relieve his homesickness and ease the transition to Toledo. The West End dinner parties seemed brighter with Florence on the scene.

With a renewed interest in salvaging his struggling Toledo operation, Libbey put advertisements in the Pittsburgh and Wheeling, West Virginia, newspapers for skilled glassmakers. His initial focus was on glassblowers and experienced foremen and line managers. Libbey went to West Virginia and Pittsburgh to interview glassmakers. From one of these trips he brought back the man that would save Libbey Glass for future generations — Michael Owens.

Michael Owens was a confident and aggressive young glassmaker. He had been moving around glasshouses in West Virginia and Ohio looking to break into management. Owens was a difficult person with strong opinions and a temper, but he was a hard-driving worker. He had been a national union officer, not what most owners would see as management material. His writing skills were minimal. His only education had come from a parish priest. A number of owners had hesitated to make him a manager because of his temper. Owens, however, was a bulldog who could achieve an objective. His leadership tool box included his fists, which allowed him to deal with difficult Irish workers. He would have made a great field infantry officer. Edward Libbey saw the potential in Mike Owens. Owens would bring problems, but he would also revolutionize how glass was made, and would to save Libbey Glass. The Libbey-Owens team would be one of the greatest business partnerships of all time.

Chapter 2

A Personal and Industrial Genealogy

The Libbey family[1] and the glass industry both came to America in the 1600s, although the early Libbeys were not glass makers. John Libbey arrived in 1634, with the British.[2] William L. Libbey (Edward D. Libbey's father) of Portsmouth, New Hampshire, entered the glassmaking firm of Jarves & Commeraise of South Boston in 1850. William L. Libbey (1827–1883) was the son of David and Betsy Libbey and had come from a line of merchants, craftsmen, and farmers. John Libbey (1602–1682) owned a trading post for British fishermen near what is today Scarborough, Maine, and developed a major plantation to augment his income. Edward D. Libbey's (1854–1925) forebears included farmers, shoemakers, and tanners. His grandfather was a tanner in Portsmouth, New Hampshire. His father started his career in the import and crockery manufacturing firm of Homer & Caswell in Chelsea, Massachusetts, the home of Edward's mother, Julia A. Miller Libbey. William and Julia had three children — Edward D. in 1854, Alice Langdon in 1858, and Sarah Miller in 1862. William's mother could trace her ancestry to Richard Warren, who was a signer of the Mayflower Compact in 1620.

William entered the firm of Jarves & Commeraise in 1850 as a "confidential" or corporate clerk to the firm's president, Deming Jarves. It was the ideal position and company for a young man to learn the glass business. William's wife's family had the business connections to land him the job. The genealogy of both the Libbey family and its company are intertwined with that of Deming Jarves. Jarves & Commeraise was truly a global company, importing and exporting glass products. It held a major stake in a number of glassworks in the New England area. The company was a great innovator under the brilliant leadership of Deming Jarves, who was the father of the American glass industry. Deming Jarves was the major investor in New England Glass and America's greatest glass manufacturer. William Libbey could have found no better person than Deming Jarves to teach him the fundamentals of the glass industry.

Jarves had entered the glass industry in 1809 with a group of investors, buying out the firm Boston Crown Glass. The company had been in operation since 1792 producing window glass. American window glass lacked the clarity of imported glass due to the lack of lead additives and skilled glass blowers. Jarves believed he could take on the

European glass imports. He and his investors added new equipment to process imported lead ore into proper additives. They hired Bristol glassmaker Thomas Cains to expand into leaded cut and art glass, which required major additions of lead. Cains taught the original German glassmakers at the company how to make leaded art glass (also known as flint glass). Jarves had studied the success of cut leaded glass manufacturer Bakewell & Company of Pittsburgh. He was a great student and took copious notes. Jarves realized he needed to import skilled glasscutters as well as blowers.

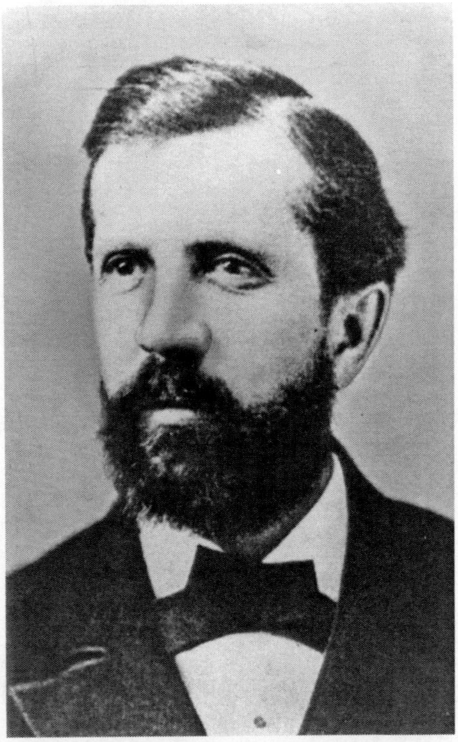

William L. Libbey, 1868. William Libbey started his career in New England in 1855 as a clerk for Deming Jarves. He became owner of New England Glass in 1867 (courtesy of University of Toledo, Ward Canaday Center, Libbey Glass Records).

Lead was the critical resource, however, that Jarves had to address if he were to be successful. The Embargo Act of 1807 and the beginning of the War of 1812 cut off the supply of lead ore. England controlled the world's only source of quality lead ore for glassmaking. It was known as red lead, which was a secondary, concentrated, processed ore. In its chemical form, it was a necessary fluxing agent. After the War of 1812, England continued economic warfare, dumping cheap glass products and keeping the price of red lead high to prevent American companies from competing in the flint glass markets. Many glass companies, including Boston Crown, went into bankruptcy. As Boston Crown failed, Jarves found new investors to run it as Boston Porcelain & Glass Company. It produced leaded art glass and tableware pieces, but lead ore shortages doomed the company. Because of the high price of imported red lead, finished American glass was more expensive than British glass.

The War of 1812 cut off New England's supply of high quality sand, which had been brought from the island of Demerara as ballast for British ships. Sand had to be of extremely high silica content and free of impurities such as iron for cut flint glass makers. Native sources were found in the Berkshires by New England makers. Pittsburgh glass houses found good sands in the Monongahela River and in New Jersey. Prior to the war, England had a monopoly on sand and red lead, but Americans proved resourceful.

The Boston Porcelain and Glass Company began in East Cambridge in 1814. On

February 16, 1818, New England Glass was incorporated, with Jarves as the principal owner. Other incorporators included Edmund Monroe, Amos Binney, and Daniel Hastings. Jarves resolved to solve the problems that had plagued the glass industry in America. He and other investors upgraded the furnaces to create a world-class operation. Jarves focused on the production of cut flint glass, which required imported red lead. Then master glassmaker Thomas Cains left to start his own glass company, leaving Jarves in a real bind.

Master glassmakers were a tight society with secret meetings and oaths. The glassmakers functioned as half guild, half union, effectively controlling the process and preventing owner inference. They never revealed their secrets. Very little was written down. Instead, practices were often made into rituals to be learned. These "protective brotherhoods" would even threaten men like Jarves who tried to discover their secrets. Jarves was banned by the brotherhood. Most European nations had made immigration to the U.S. illegal for glassmakers. The Venetians went so far as to center their glass houses on an island so glassmakers could not leave. Jarves's work with Cains had increased his knowledge, although Jarves was never part of the brotherhood itself. He was an amateur scientist who noted details and kept journals of the formulas and processing. A true capitalist, Jarves saw a huge financial opportunity in the emerging American glass industry. He was investigating a pressed lime glass substitute that would not require lead for less costly glass products such as general tableware. In addition, he was looking to discover the means of producing red lead to reduce the manufacturing costs of the more expensive cut glass products. For skills such as glass cutting, Jarves was smuggling glassmakers out of Europe. The problem of red lead would, however, have to be solved if he hoped to be fully competitive with importers. The solution would make some entrepreneur a very wealthy person. Jarves believed in his own problem-solving skills.

He had learned enough about glassmaking to start experimenting, but he needed more study of the red lead process. He had been ostracized by the glassmakers' guild and was forced to do his own research. He utilized the nearby Harvard library to study the processing of lead and hired a British engineer, James B. Barnes, to help design a roasting furnace for lead oxide. By 1819, Jarves had a roasting furnace to produce red lead from domestic sources of lead oxide, which gave him a monopoly for decades. It also allowed him to produce high-quality, leaded flint glass for art pieces at competitive prices. By using a higher lead content in art glass and other pure additives, Jarves produced an American equivalent of the famous "cristallo" glass of sixteenth century Venice. His leaded glass approached as high as 33 to 40 percent lead, and Libbey would continue to use these formulas in the twentieth century. Today most lead crystal is 24 to 28 percent lead, with the exception of Waterford, which approaches the original Jarves formula. Jarves's red lead was also a major ingredient in paint, which made Jarves's new process very profitable overall. The small village of East Cambridge on the Charles River grew with Jarves's New England Glass, as skilled craftsmen built homes.

British and German immigrants held the skilled positions, while the Irish did the menial work.

Jarves took the initial steps to industrialize glassmaking. For centuries, guild members at the master level had closely held their recipes. Glass masters frequently left firms to start competing glass houses. For example, in 1820, Superintendent Fisher left New England Glass to form Fisher & Gillerland in New York, and another New England Glass worker formed Union Flint Glass of Philadelphia. Jarves wanted to have more control over the intellectual rights of the company. American owners and investors challenged the control of the guild. Jarves started to keep a batch book of glass recipes. His batch book included the results of his own experiments and the experiences of skilled glassmakers who worked for him. He studied chemistry and hired glassmakers he could trust. At New England Glass, he controlled the batch book. This corporate practice was adopted by William and Edward Libbey. The company could no longer be held captive by a master glassmaker threatening to leave and take his formulas with him. The "metals" man or plant superintendent who mixed the glass became a trusted management employee. Jarves also tracked processes such as annealing and melting. Jarves wrested control of the glass industry from the guilds.

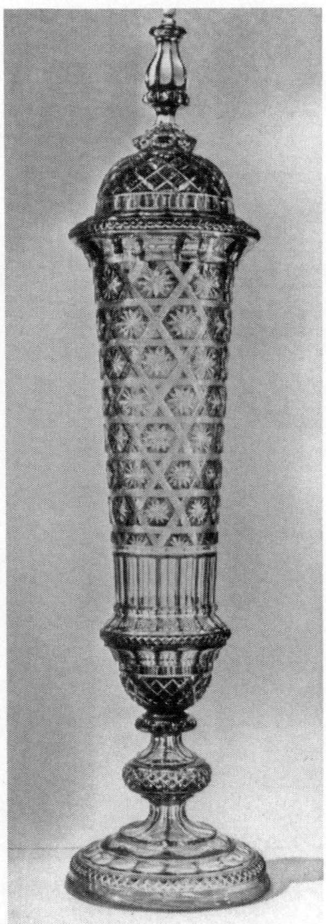

Ruby glass apothecary jar by New England Glass, 1880 (courtesy of University of Toledo, Ward Canaday Center, Libbey Glass Records).

New England Glass used wood fuel, which did not produce the high heat of the coal-fired process used to make higher quality British glass. Jarves's iconic smokestacks to increased the draft and heat generated by wood to match that of coal. Jarves also added pine tars and oils to boost the temperature of his wood fires. He smuggled Belgian glasscutters to America. By 1825, Jarves's cut flint glass was being exported to Europe. He went on to design special glass molds to improve the artwork of his pieces. In addition, he improved the pressed glass process to produce cheaper tableware pieces for the domestic market. While Jarves was not the first to produce pressed glass, Jarves's Boston Sandwich Glass Company became famous for it, and collectors refer to it as "sandwich glass." The Tariff Act of 1830 saved domestic glass operations and once again allowed for investment. New England Glass by 1830 had over 150 men producing about 25,000 pounds of glass product weekly. By 1835, Jarves had revolutionized the American glass industry, and his manufacturing company of New England Glass was world renowned.

Jarves was a leader in a group of New England industrialists who broke away from the Federalist Party to form the

Whig Party. These industrialists were followers of Kentucky Congressman and Speaker of the House, Henry Clay. Clay and his Whig Party sought the protection of America's manufacturing industries through the application of high tariffs. The glass industry was particularly sensitive to the dumping of imports, and the East Coast glass manufacturers were the most affected. In the 1820s, after George Washington, Henry Clay was the most popular figure found on a glass flask. Bakewell & Company even had flasks honoring Clay's protective "American System." Pittsburgh had its own glass protectionist in Congressman Henry Baldwin, who helped to pass the "cut glass bill," which offered protection for American cut flint glass in 1824. The cut glass tariff allowed Jarves to move into the high-end cut glass business. Without the cut glass tariff protection of 1824, New England Glass would not have been the success that it was. British, French, and Irish imports controlled the market, with Pittsburgh's Bakewell & Company having a small domestic share, until Jarves won protectionist policies from Washington. Jarves often sent special glass pieces to politicians and presidents to promote Clay's "American System." He would challenge these politicians to compare his art glass to that of the imports. In addition, the company produced commemorative Henry Clay items. William Libbey and his son Edward would continue commemorating protectionist Whigs and Republicans on glass pieces. The "American System" would help Libbey and his son well into the 1920s.

Interestingly, while the American government had placed a tariff on glass products, England made matters worse by taxing glass in the melting pot. This tax, known as the "vicious tax," lasted from 1745 to 1845. It helped destroy the ability of English glassmakers to compete in a global market. First, it added an unnecessary cost to the product. It has been estimated that this excise tax took half of the glassmakers' revenue in 1835.[3] Second, it forced the English glassmakers to lower their lead content to reduce weight and cut the tax, which boosted Jarves's quality in the marketplace. Finally, the tax forced many cut glass producers to set up shop in Ireland, where the tax did not apply. Thus, Irish glassmakers became world renowned while the English reputation declined. All in all, the "vicious tax" was a boon for New England Glass. And it helped New England Glass develop an export business.

When William Libbey entered the employ of Jarves in 1850, Jarves had major interests in New England, but his Sandwich Glass was the world's largest. Sandwich employed 300 men and boys and covered six acres. Sandwich Glass pioneered high quality pressed glass, which became known as "sandwich" or "lacy" glass. Boston and Sandwich Glass was named for the village of Sandwich, Massachusetts, at the foot of Cape Cod. Sandwich was over sixty miles from Boston. Jarves had made Sandwich a glass factory village known as Jarves Village. He built houses for the workers and brought in merchants and others to serve the glassmakers. He had personally laid out the village, designing parks and planting elms. There was a company store and independent bakeries, clothing stores, and a shoemaker. This type of communal manufacturing design would become the model for the later development of New England textile factory communities. Later, Edward D. Libbey would also look to it as a model for community design.

With its New England Glass, Sandwich's pressed glass proved the most competitive in the market place. Jarves reserved New England Glass for the higher end cut glass market, although New England Glass also made pressed glass. Jarves's pressed "lacy" glass revolutionized the American table. Jarves countered his higher costs with automation. He designed a hand press to produce pieces in a single operation instead of fusing them. He built a horse-drawn railroad to move things to and from the factory. When railroads increased shipping rates, he built an ocean steamer to ship product. Jarves's innovations had bucked the decline of American glass.

The period from 1820 to 1850 was characterized by imports, and 67 percent of those in the U.S. flint glass industry had gone bankrupt. Sandwich Glass had paid dividends of 6 percent for the period, and New England Glass had averaged around 15 percent.[4] These companies also spent $300,000 to $400,000 on capital investments. In 1850, New England Glass employed 500 and had revenues of $500,000. It was considered the largest glassworks in the world. It had a large, ten-pot furnace, with each pot holding 2,000 pounds of molten glass. Jarves was producing all four of the glass industry segments—flint, window, pressed tableware, and bottles. He had not only built a reputation for quality, but a reputation for pioneering accounting principles and management techniques. New England Glass competed successfully against British flint glass, which was subsidized by the British government until 1845.

Jarves had what would be considered today an unusual approach to the corporate entity. He believed in keeping each factory a separate and independent company. He had four major companies — New England Glass, Boston Sandwich, Mount Washington, and Cape Cod Glass. Jarves allowed these organizations to function separately with different management and product specialties. The corporation concept was just evolving, and many large organizations used a type of partnership to control different factories. Andrew Carnegie was using partnership control. Carnegie's partnership, however, managed policies of the various entities. Carnegie's approach would evolve into the world's largest corporation in 1901. Jarves's model was closer to that of J.P. Morgan, who formed trusts with interlocking directorships. The trust model allowed for independent management, but set prices and purchasing arrangements. But unlike Morgan's system of shared directors, Jarves was the single link between his companies, with the exception of his sons. George Jarves managed Mount Washington, and John Jarves managed Cape Cod. Jarves's model tended to control prices and shared technology. Jarves recognized no corporate lines for development of glass technology. Years later, Edward Libbey would model his glass trust after the corporate model of Jarves. New England Glass was always Jarves's strongest company.

New England Glass was a versatile operation producing blown and pressed glass.

It was best known for blown-molding, which Jarves had pioneered. Tableware was marketed as "blown-three-mold" glass, made from a three-piece mold, which accommodated more complex shapes. The company did art pieces but also had a mass-produced line of tavern tumblers (glasses and mugs). The most profitable production was in whale-oil lamps, which were blown and three-piece-molded. They were produced in multicolored sections such as cobalt blue and clear flint. The company produced fine cut flint glass, which could be found in the White House, had an extensive exporting business to Africa and South America. Jarves had made the company a technological leader with patents on the three-piece mold and pressing machine.

As the controlling firm's private "confidential" secretary, William Libbey learned the secrets of Jarves's approach quickly and had a good grasp of the competition in the glass industry. He had access to recipes and processing standards, making him one of the company's most trusted employees. William Libbey learned about shipping, importing, and exporting not only from Jarves, but also from his brother, Captain Henry Libbey, who was in the shipping business.

Unfortunately, 1850 was probably near the peak for New England Glass and the East Coast glass industry. The center of the American glass industry had shifted to Pittsburgh, where coal was readily available as a fuel. Coal offered higher heat and lower cost. Wood prices on the east coast were on the rise because of demand by the glass and steel industry. Pittsburgh was closer to the growing western market, which was critical for glass products, prone to breakage during shipping. Still, New England in the 1850s had the advantage of high quality sand. Berkshire sand had extremely high silica, and was free of impurities such as iron that can affect the clarity of glass. Sand quality was critical to flint and glassware manufacturers, whereas bottle manufacturers were able to live with impurities because they produced a slightly green or brown colored glass.

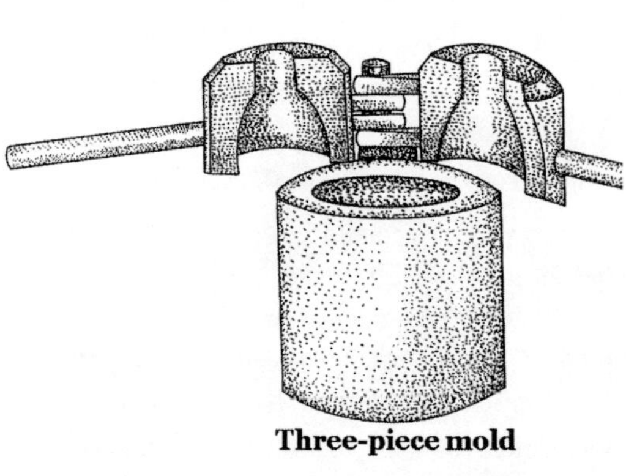

Three-piece mold (courtesy of University of Toledo, Ward Canaday Center, Libbey Glass Records).

One of Jarves's most famous "smuggled" glassmakers had a great influence on William Libbey. Thomas Leighton (1786–1849) had emigrated from Britain in 1826 to join Jarves at New England Glass. Actually, Leighton and his six sons would change the nature of glassmaking in America. Thomas Leighton, like Jarves, was interested in the chemistry of glass, and he would lead a color revolution in Victorian glass and improve pressed glass chemistry. Leighton became the superintendent ("gaffer") of New England Glass. Within a few years, five of his sons were working at the plant. Thomas Jr. was a blower of chemical glass; James was a molder; Peter and Robert were table glassblowers; and John was an assistant to his father in management.

Contrary to some myths about Libbey, it was Thomas Leighton who re-discovered the method of making ruby glass, which may have been why Jarves brought him to the U.S. Ruby-colored glass had become popular in the 1820s, but it had to be imported and re-melted at American glass houses. Ruby-colored glass had originally been developed in the Brandenburg Province of Prussia in the 1600s. Thomas Leighton in 1848 appears to have rediscovered the use of gold to create ruby glass (also known as cranberry glass). Many stories attribute the re-discovery to Andrew Long (a New England glassblower), who was said to have lost a gold ring in the batch, and later corporate legends even suggested that Libbey would secretly add gold coins to the batch every day. The fact is that a batch required about one ounce of gold per 60 pounds (much more than a single ring in a 2,000-pound pot). Also, heavy gold does not mix. The Leighton practice dissolved gold in aqua regain (a combination of nitric and hydrochloric acids) to produce a colloid of gold chloride. The development of the rich ruby color requires special heating after the piece is blown. William Libbey was in position to learn of the technique through his position of confidential secretary, but he certainly did not add gold coins secretly to the mix.

Leighton also invented mercury glass, a silver nitrate-coated glass that could be used in place of silver-plated products. Jarves smuggled another glassmaker, Rice Harris, in 1838 from England. Harris was paid $5,000 to teach Boston & Sandwich glass workers how to make opalescent glass in six months. This was popular in Victorian homes. He also taught Sandwich workers to produce milk glass, to replace porcelain in American homes. Harris was instrumental in helping Jarves make pressed glass. In addition, Jarves also used British and Irish glasscutters. With color and expert glasscutters, Jarves was able to take on the best of the glass imports.

In 1852, Jarves and Commeraise took control of Mount Washington Glass Works in South Boston. William Libbey worked through several administrative and operating positions at Mount Washington, which produced pressed glassware such as tea dishes, trays, goblets, candy dishes, and vegetable dishes. Mount Washington's product line was similar to that of Boston and Sandwich. New England Glass remained Jarves's premier glassworks, producing high quality glassware and art including oil lamps and vases. New England Glass produced major art for export such as chandeliers for churches. Mount Washington Glass, while small, was in the most competitive glassware market. By 1860, William Libbey was co-manager of Mount Washington Glass. Finally in 1866, William Libbey and master glassblower Timothy Howe purchased Mount Washington.

Jarves & Commeraise had filed for bankruptcy, and William Libbey was acting as the agent in the sale of the properties. Jarves held on to some ownership in New England Glass and some in Boston & Sandwich Glass. Libbey was ready to start his own operation and invested his savings in the Mount Washington operation. He moved the firm from utilitarian glassware into high-end art ware. He added cut flint glass pieces to the operation. Products such as decorated vases and flasks became highly successful. Business expanded beyond the capacity of the small glass house, and he purchased another operation in New Bedford, Massachusetts, moving the Mount Washington equipment to New Bedford. Libbey expanded the factory to a three-story brick building. His brother, Captain Henry Libbey, invested capital, and the company built a state-of-the-art, ten-pot furnace. The factory had shipping docks to expedite transport and minimize breakage. Henry Libbey became the plant superintendent and William handled the business end. William Libbey was an aggressive salesman with showrooms in Boston and Chicago. He traveled as far as San Francisco to sell his product.

Libbey improved the operation after making several trips to Europe to study glassmaking. At New Bedford, he introduced his fifteen-year-old son, Edward, to the glass business. The small Mount Washington Glass remained successful even as the nation entered the great recession of 1873. Competitors such as New England Glass were failing. The New England Glass board of directors offered the successful William Libbey the position of general manager in 1870, just after the death of Jarves. Libbey took a small stock position in the company after selling Mount Washington Glass for a profit of $25,000. He had established himself as a businessman and industrialist.

William Libbey tried to apply the Jarves industrial model. Jarves had developed the position of "gaffer," which was a combination foreman and glass master. The term gaffer had originally been used in England to designate the master of the glassmaking gang (head of the chair). Jarves was hoping to inject more management decision-making into the operation. Under the guild system model, a furnace pot was assigned to a glassblower and his crew (called a chair). The glassblower hired and paid the crew members, such as gatherers and annealing boys. The glassblower also set the pace and the amount of production. There was no coordination between the crews working on a furnace. Jarves and then William Libbey used the "gaffer" as the management interface, equivalent to the superintendent in other industries. The job was based on that of the European "gaffer," who acted as the glasshouse manager in the 1800s but was a guild member. Both Jarves and Libbey struggled to fully industrialize the process, but that did not occur until Edward D. Libbey succeeded in the 1890s.

William Libbey could have had no better mentor than Deming Jarves. One of the Jarves innovations, pressed glass, had created a new industry in the late 1850s. Back in the 1820s, Jarves had started the development of pressed glass. Actually, he rediscovered the techniquc of pressing glass, which could be traced back to the Romans. At the time, glassware was blown in a free form or by blowing glass into a mold. Pressed glass was made by gathering a gob of molten glass into a mold. Jarves designed and patented a machine to press glass. A lever applied pressure on a mold to make the piece. Jarves cut the cost of hand blown products in half with his early pressing operations.

Sometimes the glass appeared "foggy," but fancy patterns help hide this distortion. The pressed pieces lacked the brilliancy of a blown piece but had several advantages, the biggest advantage being cost, with the elimination of highly paid glassblowers for unskilled labor. This prompted death threats from the glassmaker guilds. Pressing also had the advantage of speed. Pressing could easily triple production rates in the 1820s. The addition of automated machinery would increase production rates exponentially in the next decade. While pressed glass didn't approach the beauty of blown and cut glass, it could be produced at a fraction of the cost. Jarves estimated that the reduced price of pressed glass caused demand for glassware to increase tenfold.[5] Jarves's prediction proved more than true. The lower price helped revolutionize the American dinner table, with pressed glass replacing wood, pewter, and pottery. Pressed oil lamps became affordable for the average family. The designs for tableware could be more intricate. The pressed glass molds lasted longer and further reduced costs. New England Glass produced beautiful oil lamps of pressed, blown, and cut glass. A new mass market and design focus led to tableware catalogs. William Libbey became an expert in the marketing of glassware.

Jarves used technology to gain competitive advantage over his competitors. Besides pressed glass, he pioneered the use of a three-piece mold. The three-piece mold allowed a deep cut pattern to be cut in a cast iron mold. Blowing into the mold created a pattern on the glass that was close to patterns cut in after blowing. It opened up a new lower price market for tumblers, decanters, pitchers, and tableware. Cut glass tableware in the 1830s cost about $3.00 per piece while mold-blown cost $1.75 per dozen. Jarves was still producing in the cut-glass market as well. A true visionary, he was a forerunner to Edward Libbey and Mike Owens. In the 1840s, Jarves brought glassmaker Joseph Magoun into the country to work at New England Glass. Magoun patented processes to eliminate mold marks in glassmaking.

Some of the glass masters that Jarves had brought into the country started to work on the improvement of pressed glass. John Hobbs (1804–1881) had worked with Jarves on the development of the pressing machine. In 1844, Hobbs and furnace engineer and Superintendent James Barnes left New England Glass to start what would be one of the nation's premier pressed glass companies — Hobbs, Brockunier & Company. The overall goal was to improve the quality enough to break into the top of the glassware market, held by cut glass companies.

A breakthrough would come in 1864 in West Virginia at Hobbs, Brockunier and Company, at the hands of John Hobbs and William Leighton. William Leighton was one of the six or seven sons of Thomas Leighton who went into the glass industry. William had moved on from New England Glass as did a number of Thomas Leighton's sons. William Leighton rediscovered the one of the earliest glasses known to man — soda-lime glass. Soda-lime glass was made from one of the original recipes for a Venetian glass known as "Crystal" (crystal). Soda-lime glass used lime to produce the high hard-

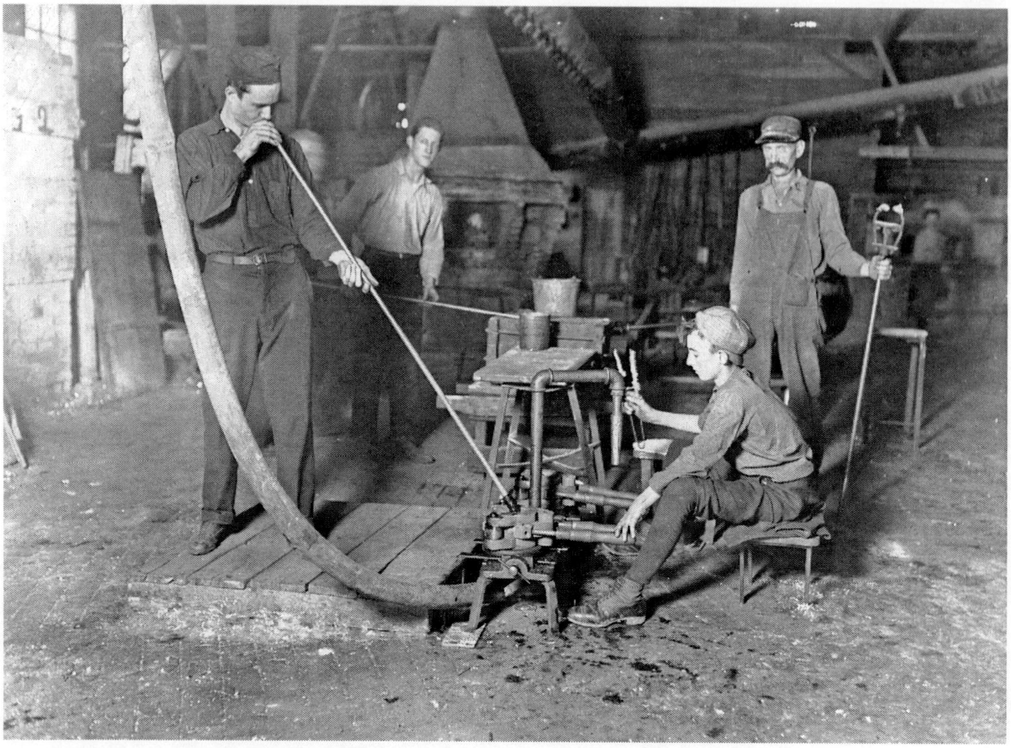

Glassblower with boy holding bottle mold (courtesy of University of Toledo, Ward Canaday Center, Libbey Glass Records).

ness needed for cut crystal glass. Soda lime not only improved the quality of pressed glass but also cut the cost by a third. The trick was experimenting to find the right formula for the pressing operation. The formula finally used was about 72 percent silica, 16 percent soda (sodium carbonate), 10 percent lime (calcium oxide), and 2 percent potash (potassium carbonate). The key to high clarity was in the purity of the silica. The Venetians had used high-purity river silica. Even small impurities would give the glass a green tint. New England Glass had discovered some extremely pure deposits of silica sand. The new soda-lime pressed glass took a big share of New England Glass's cut glass tableware market. When William Libbey became general manager of New England Glass in 1870, the East Cambridge company was rapidly losing market to the West Virginia and Pittsburgh glassmakers.

Libbey faced four major problems — a national recession known as the Panic of 1873, the cost advantages of West Virginia and Pittsburgh glassmakers, the rise of glass unions, and the innovative efforts of West Virginia glassmakers. His most pressing problem was the Panic of 1873 which had cut off bank credit. The loss of market to low-cost producers had also put New England Glass in a cash-short position. Until the 1930s, the Panic of 1873 was known as the "Great Depression." It was a perfect economic storm that consisted of a stock market crash, a scandal, and a railroad expansion bubble. The banking industry was devastated. The panic peaked in 1875, with unemployment reaching 40 percent nationally. The new oil refining business had crashed, with oil falling from $2.30 a barrel to 70 cents a barrel. Coal production just east of Pittsburgh had dropped from 45 million bushels to 20 million bushels a year. Half of the nation's railroad bonds were in default. The glass industry dropped production by 30 percent. Banker Andrew Mellon said he would never forget the soup kitchens and the homeless in the streets. Manufacturing in New England was hardest hit. New Englanders would long remember this period as the "cataclysm of the century," and "the great depression of the nineteenth century." Libbey managed to maintain the payroll through aggressive cost cutting, but the company was weakened by 1875. He suspended dividends for the first time in 36 years, which would result in a stockholders' suit against him. In 1874, Libbey brought his son Edward into full-time office employment with the company in the middle of the great panic.

As the panic subsided in 1875, New England Glass still was losing market share to western glassmakers (in Pittsburgh and West Virginia). New England Glass had used all its wood reserves by the 1860s. Oak had been favored for the annealing furnaces and high heat pine for the glass melting furnaces. The company's large furnaces could use a football field of timber every week. Realizing the problem, Jarves experimented with coal in the 1840s. By 1870, William Libbey was able to convert the glasshouse to use anthracite coal from the emerging eastern Pennsylvania fields. Being on the coast, New England Glass was able to get the coal with reasonable freight costs. Because of the conversion, Libbey's costs were similar to the costs of the western glassmakers in 1875. But then an even cheaper fuel became available to western producers. Natural gas was clean and efficient at a third of the cost of coal. This new cost advantage for the western pressed producers forced New England Glass to lose more of the utilitarian and tableware market. William Libbey increased art and cut glass production, where the higher fuel costs could be overcome.

Libbey proved brilliant at countering the penetration of pressed glass into the higher-price market. He adopted color techniques and patterns that the pressed glass manufacturers could not duplicate. One of these was the Bohemian style, which had become popular in Europe during the 1840s after being developed in Bohemia (Czechoslovakia). The technique consisted of covering a piece of glass with a thin colored flashing. Common colors were amber, green, and blue, but ruby red was the most popular. Patterns were cut or engraved into the glass, and stood out in clear glass against the colored background. The Great Exhibition of 1851 in London featured Bohemian glass in the famous Crystal Palace. Its popularity soared after the world fair. In 1856, William Libbey hired the world's most noted engraver and cutter, Louis Vaupel, who would manage his cutting department for many years to come. Limited runs of special designs and engravings such as landscapes and other scenes were beyond the reach of other companies. Jarves took the company to new heights in glassmaking, while William Libbey did the same with marketing. Libbey achieved top honors for his cut glass at the Philadelphia Centennial Exposition in 1876. Still, the New England Glass remained unprofitable because of higher fuel costs.

The western glass producers of Pittsburgh and West Virginia presented unbeatable competition. The western producers had natural gas and plentiful coal for the glass and annealing furnaces. Wages for skilled glass-workers were higher in the West, and this was causing the skilled glass workers to emigrate from New England. In addition, New England Glass's labor costs were around 52 percent, which was higher than the national average of about 42 percent. While wages for skilled workers were higher, wages for unskilled workers were lower in the West, because of cheap immigrant labor. In general, New England workers were inflexible in how they worked because of the guild system. The western producers had developed and improved pressed tableware so that it competed with the low and middle range cut glass market. The western producers were closer to western markets and further from eastern shipping, which protected them against imports. Railroads brought a ready flow of raw materials and took products to customers. These advantages were increasing with time. New England Glass's status as America's premier flint glass producer was being challenged by Pittsburgh's Bakewell Glass. The directors of the New England Glass Company stopped production in the spring of 1877 after years of heavy losses. William Libbey convinced the board to lease the property to him and continue operations.

Chapter 3

Edward Drummond Libbey

Edward Drummond Libbey was born on April 17, 1854, in Chelsea, Massachusetts, to William and Sarah (Miller) Libbey. He had two younger siblings, Alice Langdon, born in 1858, and Sarah Miller, born in 1862. Young Edward entered the Boston public school system, at the time one of the nation's best. His education contributed a great deal to his success. Boston (Cambridge) schools used the *New England Primer* and Webster's blue-backed speller in delivering an education superior to that in most of the nation. Noah Webster's *An American Selection of Lessons in Reading and Speaking* presented readings in the classics. Students were well prepared in handwriting and basic arithmetic. Theologically, New England public schools tended to be Congregational.

The public schools were heavy on Protestant theology, but the Libbey family appears to have had only a weak connection to the Congregational Church. William and Sarah joined the Newton Corner Methodist Church. When they moved to Boston, they were members of the Tremont Street Church. Some of Sarah's family had been Methodist. Edward Libbey got a strong dose of Christian principles from his grade school. Some school time was dedicated to working in the glasshouses for little or no pay. The idea was to give boys some business knowledge while supplying the labor needs of the companies. In many cases, the work would offer a career path. In many New England towns and villages, school was divided into a morning and afternoon session to facilitate boys working a full shift in the glasshouses. At Sandwich Glass, boys would receive a gift of firecrackers for the Fourth of July. Community cooperation wasn't limited to the school schedule; when the factory needed a water channel cut, the citizens volunteered to help in the digging. The glass factory and community had interrelated missions and a symbiotic relationship in a type of arrangement going back to the middle ages.

Edward D. Libbey was sent to Maine's prestigious Kents Hill Academy (known then as Maine Wesleyan Seminary) in 1872. Kents Hill was one of the nation's best preparatory schools. It had a long list of wealthy and famous graduates even in 1872. Edward's mother, Sarah, hoped that he would become a Methodist minister or attend

Harvard or Boston College. Edward had received full membership in the Methodist Church in September 1870. He had been a good student all his life. Records show he graduated with honors with courses in Latin, Greek, rhetoric, arts, and philosophy. Kents Hill had an outstanding series of commercial courses which Edward enjoyed.

In 1874, a throat infection disabled his voice; a career in the ministry was doubtful. This would also have put him at a disadvantage in the rhetoric-driven colleges of the time. However some evidence suggests this may have been a ploy to return to his father's business. He returned home seeking a job in the business. William Libbey still hoped that his son would go to college, and persuaded him to take some courses at Boston College. Edward, however, showed little interest. Still, his education was superior to that of most Americans.

Edward D. Libbey joined his father at New England Glass in 1874 at the age of twenty. William was a tough master, requiring Edward to work the lowest clerical positions. He learned the work of every office employee in this difficult apprenticeship. His main interest was the same as his father's — marketing and product development. He had one interesting opportunity during this time — a trip to Wordsley and Stourbridge, England, to study the process of needle etching.[1]

William had employed a British agent in 1879, and they corresponded weekly on developments. W. H. Richardson had developed a lathe to needle etch products through a wax coating. Edward befriended the etch master, A.J. Nash, in England and learned the process. On his return, Edward helped the company develop the technique in the United States. This etching technique would be applied to lamp globe and shade production and would be a highly profitable branch of the business. It also proved to save labor. The lathe needle outlined the relief and removed a thin line to the glass surface. The product could then be dipped in acid for etching. Hand etching a single globe or shade could take weeks, while needle etching took the process to several days. New England Glass even applied it to tableware production. It was the first of many accomplishments of Edward Libbey as a glassmaker.

Young boys in New England often worked in a glasshouse after school. Edward always enjoyed the office work. He started at the age of twenty as a clerk under his father's direction, but he loved the art of glass. He had the passion of a collector. His second important business trip was with his father to the Centennial Exposition of 1876 in Philadelphia. New England Glass, Mount Washington Glass, and Boston & Sandwich were some of many glass companies exhibiting there. The Philadelphia Exposition celebrated the 100th anniversary of the Declaration of Independence. Nearly 10 million people, or over one-fifth of the U.S. population, would visit this summer world fair. It highlighted the growing dominance of American industry. The most visited exhibition was that of Gillinder & Sons, who built an operating glasshouse on the fairgrounds. They demonstrated pressing glass, producing souvenirs for sale. They also introduced a type of frosted glass using hydrofluoric acid. Their wildly popular Lincoln bust, Cinderella slipper, and lion paperweight, produced in clear, frosted, pink, and red, made the company a large return on its investment. Gillinder & Sons took most of the gold medals and attracted most of the crowds. The success of Gillinder & Sons inspired Edward Libbey in his 1893 World Fair exhibit. Mount Washington Glass exhibited a 48-foot high crystal fountain at the fair. New England Glass had a smaller but still impressive exhibit in 1876. It featured a Daniel Webster cut glass punch bowl. The company's apple and pear paperweights were popular as souvenirs. The bigger success of the Centennial Exposition was that it started a trend in deep cut flint glass that swept America, reaching its zenith in the 1890s.

William and Edward Libbey spent days at the fair finding inspiration. The big attraction was the right arm and torch of the Statue of Liberty, which could be visited for 50 cents. For another 50 cents, the visitor could purchase the most popular souvenir of the fair, a glass arm and torch, which remains popular with collectors to this day. Edward was able to view exhibits by 130 glass firms. He also studied new inventions — the telephone, typewriter, automatic screw machine, and the electric dynamo. He probably unknowingly passed other budding industrialists such as Thomas Edison, George Westinghouse, George Pullman, and H.J. Heinz. He would later note the importance of this experience in developing his business and marketing strategy.

From 1874 to 1877, Edward Libbey continued to cycle through clerical jobs at the company. The deepening of the recession following the Panic of 1873 eventually caused the company to dose temporarily in 1877. The furnaces were banked on April 12. Efforts to sell the plant failed as the recession deepened. William Libbey took over, leased the plant for operation, and resumed production in March of 1878. Employment had dropped from a high of 500 to 100 in 1877. Edward Libbey suffered a personal setback with the death of his mother in July of 1879. His father would remarry in 1881, and Edward appears to have had a close relationship with his stepmother, Sarah C. Libbey. She was well educated and had traveled extensively in Europe.

William Libbey, while not a glassmaker, learned the trade as his mentor Deming Jarves had. He developed and patented a new process for making cameo glass. His plan to return the company to profitability focused on an aggressive marketing strategy, leaning heavily on development of colored glass products. In particular, he hoped to add beautiful colors to flint (leaded) glass, novel and unique colors that could be layered into complex cut glass. The Centennial Exposition had proven the popularity of colored glass and new products. In particular, William Libbey hoped to revive the ruby-red glass (also known as cranberry glass) so popular in Europe, as well as to develop color variants. He initially assigned several of Leighton's sons and Andrew Long to the project. Libbey hoped for shades of amber, pink, and peach. Experiments and failures were endless. The company continued to lose money as the country slowly came out of the recession. But William was successful at expanding the high-end cut glass market. By 1880, William Libbey brought his son fully into the company and renamed it William L. Libbey and Son. With the change, Edward became more involved in the marketing and sales functions.

In early 1882, they added some key craftsmen to aid their marketing strategy. The most important of these was Joseph Locke, an English glass designer. Locke had extensive experience in glass chemistry and coloring. Another was skilled glasscutter William C. Anderson. Anderson joined William Donovan in the cutting department. Deep cut pattern flint glass was becoming a major moneymaker for Libbey Glass, as the color project floundered initially. Libbey was having much success selling cut bowls in the $25 price range (about $500 today). His was the premier cut glass company in America, and his flint glass was considered the best in the world. Some of this is attributable to the company's focus on glass chemistry. Libbey's formulas called for higher lead contents (22 percent to 30 percent) than those of his competitors, giving it more weight and clarity. Libbey invested in the best cutters and engravers he could get from Europe. His strategy was to dominate the world market in quality, and this required him to pay world-class wages. This had been the strategy of his father and Deming Jarves. Libbey, developed a powerful marketing plan, his forte. Contributing to Libbeys success was the sales skill of Jefferson D. Robinson, who had started as a young boy with Edward. Robinson and Edward Libbey had become close friends over the years. Both had a strong belief that they could develop a market for colored flint glass.

The tradition of fine flint glass production in New England went back to Deming Jarves and the founding of New England Glass. The name "flint" originated with seventeenth-century English glassmakers, who made cut crystal glass from pulverizing silica flints. In 1673, George Ravenscroft pioneered the formula for leaded flint glass. Lead oxide in amounts from 20 percent produced clarity and a crystal-like quality. Potash (potassium carbonate) was added to crushed silica to improve clarity and help the fusion of the melt. Jarves took Ravenscroft's early formulas and improved on them, using the pure Berkshire sand of New England. Jarves's own red lead production added to the purity of the flint glass of New England Glass. The clay for the furnace pots was imported from Europe to assure purity. Over the years, the company further honed the melting, heating, annealing, and cutting practices. By the 1870s, there were no competitors in the league of New England Glass. Still, cheaper pressed glass was cutting into the tableware market. Pressed glass sold at about a third the cost of flint glass. In addition, soda-lime glass used for pressing was easier to color. The cost of pressed soda-lime tableware made it popular with the middle class. This popularity was a major factor in the downfall of many famous cut glass shops such as Bakewell Glass in Pittsburgh in 1882. However, leaded flint glass remained the standard for quality, cutting, brilliance, and clarity. New England Glass managed to weather the storm for a few more years.

William Libbey took on the production of etched ruby and amber glass as a personal project. This beautiful glass had been popular in the Middle Ages and was making a comeback in Germany. In 1879, he hired W. Rice Harris of Birmingham, England, the son of Rice Harris, who had worked for Deming Jarves in the 1830s. W. Rice Harris hired European glassmakers for New England Glass. They got passage to America and housing. But New England Glass had trouble keeping immigrant glassmakers. William Libbey, like his mentor Deming Jarves, collected and read every book he could find on ruby glass. Edward Libbey got involved and it soon became a personal quest for him as well. Libbey kept good records in his batch book.

In the summer of 1882, the first successful amber flint glass experiments were completed. What was produced was a variation of the work produced by Thomas Leighton years earlier. The final piece of the puzzle came from Joseph Locke, who had been hired by Edward. Locke filed for and received a patent on July 24, 1883, for heat sensitive colored glass, which would be known as Amberina. The secret was not only the addition of gold, but re-heating and annealing the finished piece to maintain the perfect color. Re-heating the blown piece in the glory hole developed most of the color. The glory hole was in the side of the melting furnace, or as in the case of New England Glass, a separate re-heat furnace. By the 1850s, the "glory hole" generally referred to a small separate furnace used for re-heating and fire polishing. The glassblower would need to reheat the pieces a number of times while forming continued. It appears a number of companies had discovered the process, but because it was merely a manipulation of the process, it could not be patented. William and Edward thought differently.

The Locke process could be used to create an array of red, orange, and yellow colors. Locke went on to patent these color variants — Pomona in 1885, Wild Rose in 1886, Agata in 1887, and Maize in 1888. They Libbeys took Jarves's industrialization a step further with patented formulas. The patent was issued to the inventor, but the company maintained the rights to produce the product. This arrangement fostered employee creativity while protecting the corporate investors. Where the effort involved many, Edward Libbey was issued the patent under his name.

William Libbey died in August 1883, before the amber glass went to market. Edward took over the company and took possession of the batch book. Libbey went a step further, patenting cut designs by his skilled cutter, William Anderson, such as Hobstar and Middlesex in 1887, and Russian a year later. Seeing the success of creativity in sales, Edward offered prizes for employee designs. Anderson's first pattern won the first prize. The prizes and patents helped ease resistance of the guild members to the concept of intellectual rights. He actually won the support of the guild that Deming Jarves had failed to get.

The success of Libbey's Pomona and Amberina glass is only recently coming to light. It took chemistry, cost reduction, processing, and marketing. Libbey's old batch book has the Pomona formula under "XXX Pomona." It was a colorless glass of half the usual lead content, which substantially reduced the cost. The color and quality were developed through acid etching and staining. While Locke may be considered the inventor, these were processes Edward Libbey had studied in Europe. Still, science and technology were only part of the story. Libbey created a market, and he had the genius to bring the market and technology together.

Edward and his young salesmen Solon O. Richardson, Jr., and Jefferson D. Robinson took the product to the market. Robinson was less enthusiastic, believing the market for ruby type glass was not large. The first new color was Amberina, which was a pale yellow shading to ruby. Edward Libbey coined the name. Blown mold pitchers, vases, and other tableware were the first products and were all cut flint glass for the high-end market. One of the first pieces can still be seen in the Toledo Museum of Art. Libbey proved to be the best salesman, landing a large order for the Amberina line at Tiffany in New York. Locke and Libbey followed quickly with blown mold patterns at a lower price, which were even carried by department stores. Some of the pressed glassmakers tried to copy it. Libbey aggressively defended his patents, forcing competitors to pay license fees or get out of the business. Hobbs, Brockunier & Company purchased a license to make Amberina in pressed glass. Mt. Washington Glass was forced to stop production of Amber Rose. It was an important lesson for young Libbey that innovation could make money in different ways. Competition wasn't the only path for glassmakers; cooperation with competitors could pay dividends. It was a very different approach in an industry where cut-throat competition was the norm. Most glassmakers avoided patents to keep formulas totally secret, limiting access within the companies to the formulas. Licensing was a new business model that Libbey would take advantage of many times. But he was less successful in his efforts to prevent copying of his cut designs.

Locke produced other color variants such as Peach Blow and Agate. These colors proved as popular as his original Amberina. Variants of Peach Blow were made by a number of companies, but its origin goes back to the work of Thomas Leighton, who in 1883 had moved on to Hobbs, Brockunier & Company.[2] New England Glass's Peach Blow shaded from creamy white to a deep rose. Locke also developed an etched and tinted product of pale amber or strawberry known as Pomona. These products soon were being called "art glass." Thanks to the art glass, in 1885, Libbey reported his first profit since 1873. But that profit was threatened by the cheaper pressed glass copies coming out of West Virginia. The innovation battle between Joseph Locke of New England Glass and Thomas Leighton of Hobbs, Brockunier & Company in West Virginia continued for many years. For his part, Libbey added the art colors to lamps, which expanded on the success. Color art glass of the period consisted of five types: one solid color, heat-sensitive shaded one color, overlaid colored glass, a flashed thin color, and a chemical stained product. The last three types could be engraved with dramatic designs. Libbey was producing all five types.

Libbey primarily made flint cut glass or blown, pattern-molded glass as opposed to the pressed glass. Libbey's main competitors in cut glass and high-end product in the 1880s were Bakewell, Gillinder and Sons, Mount Washington Glass, Boston and Sandwich Company, and Christian Dorflinger. The pressed soda-lime product competed on the lower end of the market. Hobbs, Brockunier & Company was the major pressed glassmaker. Libbey did have a line of pressed glass as well as using pressed glass parts in his lamps, but with his high fuel costs, the pressed glass tableware market was unprofitable for him. Libbey was able to capitalize on the popularity of art paperweights, and to a lesser degree, on the growing demand for kerosene lamp chimneys. His interest was always with art glass, which he considered the "trade of gentlemen" and the work of artists. Libbey Glass worked with the guild and blower's union to support blown product over pressed. Locke developed a unique two-step process for patterns. First, a pattern mold produced a blown patterned piece. The piece was reblown into a blank piece and mold, which embedded the pattern, leaving a smooth surface. Libbey's emphasis on the art blown product won him long-term support with many in the guild and union.

Libbey's early success was rooted in his ability to combine art and manufacture. He went further than his father or Deming Jarves in coordinating sales and manufacturing. Today we would call this marketing. Like dress designers today and the automotive designers in later years, Libbey moved to annual design releases. He combined new colors and patterns to create lines of tableware patterns. Libbey then used novel approaches to promote and advertise his product. He was not the first to do this, but his coordinated approach throughout the business was unique. The idea of a coordinated glass table service went back to London retailers in 1780. By the 1820s, table service included an array of glasses and beakers. Libbey not only expanded the product but its use in society. He sold creativity. In one respect, he brought craftsmanship and art into the industrial age. He supported this type of marketing by strengthening his sales offices and printing improved catalogs. He sent his friend Jefferson Robinson to open a Chicago sales office, in addition to the New York and Boston sales offices. But his success in marketing was offset by spiraling production costs due to labor and fuel.

The very nature of the glass industry was changing in the 1880s. The community-based industrial model was in a state of flux. The iron and steel industry was rapidly changing to a new industrial model driven by technological changes. Mass production was replacing craftsmanship. Automation allowed unskilled machine operators in iron and steel to replace skilled workers. The shift created the union movement in the United States which was widespread even in the crafts industries in the 1880s. The Knights of Labor was a national union that had gained ground in the glass industry.

Glass production in 1880, however, was basically the same as it had been for

2,000 years. A sixteenth-century Venetian glassmaker would have been right at home in Libbey's glasshouse. A Libbey Glass historian put it this way:

> Strange as it may seem, if a master workman of Libbey Glass Company were to be magically transported today to a Venetian glass house of several centuries ago, he would find at hand most of the tools he had always employed, and he would call them by names that would be perfectly understood by his Italian fellow workmen.[3]

The glass industry was still labor intensive, but skilled workers — 20 to 30 percent of the workforce — controlled the operation.[4] Labor costs amounted to 40 to 50 percent of the manufacturing costs.

Special clay pots were charged with flint, silica sand, metal oxides, lime, soda, and cullet depending on the type of glass being made. The mix was known as a batch. Cullet was broken glass which facilitated the melting process. Glass melts at a much lower temperature than the individual ingredients. The following is a Jarves formula for European Crystal.

1,200 pounds of pure silica sand
800 pounds of red lead
400 pounds of potash (potassium carbonate) for fusion and melting
50 pounds of nitre (known saltpeter or sodium nitrate)
10 pounds of lime
10 ounces of antimony oxide (for clarity)
24 ounces of manganese oxide known as glassmaker's soap (it was a decoloring agent)
32 ounces of arsenic (added to improve clarity)
20 ounces of borax[5] (a fusing agent to lower melt temperature)
Metal oxides were added for color such as copper for green and cobalt for blue.

The trusted gaffer or the owner oversaw mixing, but unskilled workers or boys did the actual shoveling and mixing. These boys lacked the chemical knowledge to understand the mix, and this helped keep the secret of the process. Libbey maintained the only batch book but might detail the particular mix for the gaffer to oversee. Libbey's early batch books were in his own handwriting. Skilled workers made the pots or crucibles. The pot maker was at the skill level of a glassblower, since pots were critical to glass quality. The clay was fireclay, which flint glassmakers such as Libbey imported from Germany. High purity and low reactivity were required of pots in the flint lead glass industry. The melting pot clay was mixed by barefoot workers and then formed and air-dried. One problem in pot making arose in the flint industry after the switch from wood to coal. Pots or crucibles had had open tops for centuries, but coal dust caused the lead in flint glass to precipitate out, destroying the melt, so covered pots were required. A pot in the 1880s held between 500 and 3,500 pounds of melted glass. Melting pots would be put in the glass furnaces, which held eight to twelve pots. Melting pots were pre-heated before the charge mix was added slowly over a period of hours. This pre-heating was known as "fritting." Fritting fused the glass and prevented thermal shock from breaking the pot. The furnace would heat the pot to 2,500 degrees Fahrenheit over a ten- to sixteen-hour period. Boys fueled the furnace.

Production revolved around the individual pot. A crew (also known as a gang, shop, or chair) worked an individual pot of glass. The term "chair" was used in Europe, referring to the bench from which the glassblower worked. The large domed furnace had a working door for each pot of glass, known as a Bocca. A gang was assigned a Bocca and pot. When a pot was ready, a boy, known as a "knocker-upper" or "call boy," went from house to house in the town to notify the crew, particularly the early morning crew. The nature of the working crew varied with the type of glass and product. The crew or gang or shop in Libbey's blown art glass factory was a five to six man team that worked on one of the pots. The gang worked two shifts (moves) a day; one from 1 A.M. to 6 A.M. and another from 1 P.M. to 6 P.M. Most plants closed over the weekend while boys maintained the furnace heat.

A master glassblower (sometimes called a gaffer) was in charge of the gang. The term "gaffer" is said to be a corruption of the word grandfather. In Libbey's glasshouse, the glassblower usually had an assistant or helper. The next skilled job in the gang was the gatherer. The gatherer was considered a unique craftsman with his own guild, pay rates, and apprentice systems. A gatherer used the blowpipe to gather up a pasty ball of glass ("gather") from the pot. A gatherer might also have an apprentice or journeyman in training to help. Management and the union contended over the use of apprentices and assistants versus boys. The gatherer or a servitor formed the mass on an iron table known as a marver and worked the mass into a viscous ball to turn over to the glassblower.

Both the gatherer and blower would have apprentices who did general labor for the crew. The blower formed the product at the end of the blowpipe either by free forming or blowing into a mold. The blower moistened his mouth so that steam would help in forming the bubble at the end of the blowpipe. Free-forming required the most skill. If the piece cooled too much while he was working, the blower handed it to a middle-boy to reheat it in the glory hole. After the blown shape was formed, the piece was transferred to a pontil rod, where the blower would shape it with wooden or metal tools. As the piece was being formed, a bit-boy might bring small pieces of glass to add things like handles. Rims had to be reheated in the glory hole to become fully smooth. If other work was needed, a warming-in boy reheated the piece in the glory hole. The piece was then snapped off the pontil and broken off the rod, leaving the well-known pontil mark. If the piece was blown in a mold, a mold boy was needed to coat and release the mold. In the case of pressed glass, the presser was the highest-paid craftsman. It usually took five to six hours for the gang to fully work a pot of glass. Gangs were usually divided into morning shifts and afternoon shifts. Outside of the hot work were skilled cutters and engravers. Because of the heat, glasshouses closed during the summer months, although this was often a point of contention between the union and management.

Boys did the least skilled work of handling and moving material. Usually boys fueled the furnace to ready melted glass for the first day shift of the gangs. In addition, boys helped the gangs. Boys cleaned and prepared molds (mold-boys) for the glass-blower. An in-boy (also known as a taker-in boy) took the blower's finished product to the annealing furnace (known as the lehr or leer), and an out-boy (taker out) took it out of the annealing furnace. Annealing was a critical step of forming and cooling. Without annealing, a piece might literally explode on cooling. Poor annealing might cause the piece to break during cutting or finishing operations. Boys took the pieces hot to annealing where the cooling was controlled and slowed to an eight- to ten-hour cycle; then boys unloaded the pieces. Boys cleaned blowpipes and other tools as well. The boys worked twelve-hour shifts, working production or maintaining the furnaces through the night. Boys accounted for 15 to 25 percent of the total workforce. Management strove to use boys on the crew instead of higher-paid assistants in training. Low-paid women might be used in the packaging and shipping department. Women were also utilized in cleaning, washing, and the final polishing of product. Interestingly, Jesup Scott's manual arts university in Toledo specialized in the training of women for industry. In the 1870s, the community often supplied boys for "business training," but as labor became scarce, boys were paid a daily wage of around fifty cents. This started to add to Libbey's labor costs in the 1880s.

The 1880s saw the rise of the glass unions, particularly in the West where the steel industry was unionized. From 1850 to 1880, Pittsburgh was America's glass city, with 44 glasshouses producing nearly a third of the nation's output. Unionization of glass workers presented problems for both the workers and owners. It was a complex movement in which Libbey's company found itself in 1882. The general union movement in the industry progressed in industry segments such as window, bottle, and flint glass; organizing by craft also was offered to the blowers and gatherers. The alignment by industry segment made more economic sense to the union organizers, but the craftsmen, seeing themselves as artisans, preferred a crafts organization like the century-old guilds. The guild issue was bigger in New England than the wage issue. Glassmakers were some of the highest paid craftsmen in America. The Knights of Labor scaled wages by trade and included all the crew. The Knights found they could gain ground in the glass industry by modifying their approach and allowing crafts specialization. The glass industry of the 1880s was little affected by automation, unlike the iron and steel industries.

New England Glass was one of the earliest glass companies to use factory management system. Edward Libbey had fully implemented what his father had started.

There was a plant superintendent (still referred to as the gaffer) or floor manager. The furnaces had a shift foreman. Large departments such as cutting and etching had a department superintendent and a shift foreman. On the hot side of the operation, the industrial management structure clashed with the old guild structure. This created a factory floor struggle between the master craftsmen such as the glassblower and the management foreman. Under the factory system, foremen were challenging the management role of the head glassblower, who under the guild system hired, paid, and set the production schedule and work pace. The biggest challenge was the work pace, which could directly affect labor costs. The New England Glass structure was new; and while there was a plant superintendent, Edward Libbey had the supervisory role of a plant manager today. In addition, competing national unions complicated the situation. Unionization was not easy in the glass industry because of its strong guild heritage. Guilds were organized for blowers, gatherers, and other positions separately. Initially in New England, the blowers and gatherers had formed their own unions. A master craftsman such as a master blower, until the late 1870s used an operative, like a sub-contractor today. The gaffer or master blower hired his gang and paid them. The gaffer also controlled the pace of the piece production. The new industrial model changed the lines of control.

In the cutting departments, Edward Libbey had extended the industrial model of his father. Designers and artists filled the superintendent and foremen positions and were usually master cutters. He also had artists and craftsmen in non-management positions. In the crafts model, the artist designed, cut, and finished one piece of work. He would have journeymen and apprentices to help. In Libbey's industrial model, the designer would stretch or etch the design on the glass piece. It would then be passed to the rougher, who used a steel wheel and sand to cut the pattern. It was then passed to the smoother, who deepened the cuts and added fine cuts using a wheel known as a *craigleith*. Finally, the polisher used rouge and tin oxide on a walrus-hide lapped wheel. This approach increased production without degrading the craftsmen. The wheels could be driven by overhead drive belts in an assembly line row. Some special pieces might require the full-time work of a cutter. Libbey's famous McKinley punch bowl set took four weeks' work by two master cutters. Cutters were considered the master craftsmen of the factory, who had progressed through rougher, smoother, and polisher. In shops where the union took over, it lumped the roughers, smoothers, and polishers together as one trade. Another artisan was the engraver, who used a single lathe. Applying linseed oil and emery powders to a copper wheel during engraving, the engraver did everything from detailed pictures to monograms.

In the late 1870s, local unions started to join national unions. These tended to be aligned with industry segments. At a local level the unions had lodges which covered the needs of the individual craft. In the flint glass industry, the blowers were the highest paid. Typical wages were around $3 a day, around $2.50 a day for gatherers,

$2.50 a day for engravers, and $2 a day for mold makers. European glassmakers made less than half of these wages. Individual lodges established the crafts apprentice program and restricted membership. The national movement started around 1875. The window glassmakers joined the Knights of Labor in 1878. In 1858, the glass blowers took the first step from guild to union in the formation of the Glass Blowers' Benevolent Society. The Pittsburgh flint glassworkers came together to form the American Flint Glass Workers' Union of North America (AFGWU) in 1878, which by 1882 had spread to West Virginia and west, and finally spread to New England. Libbey's New England Glass joined the AFGWU in 1884. The AFGWU was truly a crafts union representing only the skilled workers, or roughly 30 percent of the glass industry workforce. Boys, in particular, were left out of union protection and bargaining.

In 1885, the AFGWU started to centralize its control over local unions. Libbey faced a short strike in 1885 over wages, and to some degree, recognition of the AFGWU. Libbey accepted the authority of the AFGWU but never settled the wage issue. He opposed union demands to control hiring, work hours, and the amount of production. The AFGWU had a regional wage scale that it wanted to be made uniform. The highly profitable western (Ohio, West Virginia, Indiana, Kentucky, and western Pennsylvania) glasshouses, with cheap fuel, paid higher wages and required less output of the gang. The western workers made anywhere from 15 to 50 percent more than eastern workers.[6] Boston & Sandwich Glass Company had been the front line of the dispute. The union had organized there first, in 1879. A strike over equalizing wages with the west began in January of 1887, which essentially ended the company. It was an end that Libbey and his workers were trying to avoid, but the wage issue was central to nationalizing the union.

Still, glassworkers were higher paid than other American craftsmen, and there is extensive evidence that on labor sensitive products, even the western producers were not competitive with foreign products.[7] The average daily wage in 1888 was a $1.40. The average skilled glass worker made $3.30 per day compared to $2.30 for a carpenter, $3 for a bricklayer, and $2.30 for a blacksmith. Furthermore, glass industry wages had shown a significant increase from 1840 to 1880. The average American factory worker made $1.20 per day, about the same as a starting or unskilled worker in the glass industry. Once an employee became a skilled worker, he progressed through apprentice and journeyman to master. There was a great deal of security and geographic opportunity. Only the highest-paid, skilled, tonnage steel workers, such as rollers, were paid higher. The pay allowed glassworkers to be middle class, like merchants. Skilled glassworkers tended to own their own homes and had status in society. They were seen as craftsmen and artists, not factory workers. This was especially true in the cut flint glass industry. They had surplus income to spend on luxuries such as the cut glass they produced. The workers were pulled in different directions by industrialization and unionization.

Libbey viewed them as craftsmen and artists, just as he viewed the finished products as art. Libbey and the Bakewells of Pittsburgh were considered the industry's best employers. He had the utmost respect for his workers. The workers were sympathetic towards Libbey's need for lower costs, but the pressure was coming from the national committee of the AFGWU. With rising fuel costs, Libbey was in no position to meet western wages, although he did give a small increase.

Another problem was the establishment of work rules by the AFGWU, which took away the flexibility of owners and increased costs. In 1879, the AFGWU banned overtime by limiting the number of turns a gang could work. This forced many companies to add gangs. In 1880, the union set limits on lower paid apprentices and required summer shutdowns to be a minimum of six weeks. The union defined the crew size, restricting the tendency of management to substitute boys for skilled labor. With cooler summers in New England Libbey felt he could operate longer. He could ill afford these restrictions and did not abuse employees. Even Libbey's treatment of boys, who were not covered by the union, was fair and humane. The men respected his efforts and tried to walk a fine line between their national representatives and his management. He had been successful at keeping the whole operation going under difficult circumstances.

Colored art glass carried the Libbey Company until late 1886 when fuel availability became a problem again, along with union wage demands. By 1888, Libbey was again losing money. The decline in profits was attributable to the penetration of pressed glass into the art glass market, continuing high fuel costs, further cost reductions by the western producers using cheaper and more efficient natural gas, and labor problems. The labor problems went beyond union demands for higher wages. Historically, glassmakers had always used the leftover molten glass ("metal") to make "fancies," articles of glass made for the glassworkers. Generally, these were glass canes and paperweights for personal use or display. Libbey's East Cambridge plant had taken to making quantities of valuable pieces for resale. It was even reported that finished glassware was taken from the warehouse for resale.[8]

The use of natural gas in the west presented a real problem for Libbey. Gas was even cheaper than coal and it supplied a higher and more consistent heat. Natural gas offered other advantages for flint glass makers. At times, coal caused the leaded glass to pick up carbon, which in turn caused a precipitation of lead, destroying the clarity of the glass. Gas firing significantly extended the life of the furnace brickwork and the clay pots, which again was a major process cost reduction. The cost advantage of the western producers widened in the early 1880s as Pittsburgh and West Virginia glasshouses switched to natural gas. George Westinghouse was pioneering the transmission of natural gas over long distances in Pennsylvania. By the mid–1880s, the gas boom in Ohio created a large number of new, low-cost producers. Cheap fuel created an industry boom, with over a hundred new companies created in two years. While the Ohio gas boom was short-lived, it brought a large number of new art and flint glass makers into the industry between 1886 and 1889.

Another factor was that the glass market had shifted west. Libbey's shipping costs were double those of West Virginia and Pittsburgh to Chicago. Pittsburgh, in particular, was a rail transportation hub for the United States. While New England only had two glasshouses left, Pittsburgh had 46 and West Virginia over 20. The new Ohio glass companies found themselves in the middle of the market growth.

Libbey remained determined to find ways to stay in New England in 1887, following his successes with colored glass. He tried to compete in the lamp chimney segment of the flint glass market. Chimneys were hand blown at the time, and the business was booming. The discovery of oil and its distillation to kerosene in 1859 brought a new demand for kerosene lamps. Libbey was very competitive in the high-end art lamps, but the booming market was in the everyday, utilitarian kerosene lamp. The going price was at about 30 cents per dozen chimneys, and Libbey's costs in 1887 were about 29 cents per dozen prior to shipping west. Pittsburgh glasshouses and the new Ohio glasshouses were expanding rapidly into the production of chimneys for kerosene lamps, but even these were under pressure from imports. Lamp chimney production was highly labor sensitive, and even with a 45 percent tariff on the incoming value, cheap European labor allowed foreign manufacturers to undercut the American producers.[9]

At the same time, Libbey's versatile salesman, Solon O. Richardson, was working with Thomas Edison on the possibility of mass-producing the new electric light bulb. New England Glass was the major supplier of glass tubing, which Edison had initially used to make light bulbs. New England Glass had also been Edison's supplier of chemical glassware for his lab. Richardson did land some light bulb business for special, hand-blown sizes. New England Glass had some orders with smaller electric firms due to its experience with Edison. Richardson broadened into the operational side of the business and was able to coordinate new products such as light bulbs. While in his late twenties, Richardson became an important advisor to Libbey, although they often disagreed. Richardson was now the secretary of the company and was pushing for Libbey to get involved in bulb production. Libbey lost the business to Corning Glass of New York in the early 1880s. It was a non-union shop with low labor costs. Corning used a fuel mix of wood and coal.

As Libbey struggled with lost market share, high fuel costs, and price competition, the union threatened a strike for higher wages. Libbey found himself negotiating with the national union, which was demanding wage parity with the west. Libbey referred to the national committee as the "wrecking squad." He had an ally in James "Gaffer" Smith, who represented the eastern locals, but in the national convention of 1887, the union officers overruled Smith. Western representatives controlled the national union. Smith was said to have lost the wage debate with an up and coming union official, Michael Owens of West Virginia. Some say that Mike Owens was sent to New England to help the East Cambridge workers stand against the company. It appears unlikely that Mike Owens was directly involved in the negotiations in New England, but his debate victory at the 1887 convention did contribute to Libbey's decision to look west for the future of his company. Still, the major reason for Libbey's decision to move was the competitive advantage of natural gas.

Chapter 4

Toledo and Northwest Ohio of the 1880s

Northwest Ohio of the 1880s was the nation's boom area. The discovery of a pocket of natural gas known as the Trenton deposit that ran through Northwest Ohio and Indiana created an industrial boom that fostered a rush of glass companies to Northwest Ohio. Gas seemed so abundant that towns offered free gas to lure companies to locate there. Sleepy little villages such as Fostoria, Tiffin, North Baltimore, Bowling Green, and Findlay became boomtowns overnight. In 1880, there were no glasshouses in Northwest Ohio. Within a 25-mile radius of Findlay over one hundred glass companies opened from 1886 to 1890. Fostoria had 15 glasshouses, Bowling Green had 11 glasshouses, and Tiffin had 8 glasshouses. Thirty-one glass factories were built in Findlay, Ohio, alone from June 1886 to June 1887![1] Among this migration would be Edward D. Libbey; however, Libbey would break with other glassmakers in selecting Toledo, which at the time did not have a single glasshouse and was twenty miles out of the main pocket of natural gas. Industrialization was so rapid in Northwest Ohio that a labor shortage was created. Towns advertised seeking large families. Trains brought young orphans from big eastern cities to these Ohio towns to supply workers. In Toledo, young women were enticed to enter the glasshouses. Ohio cities advertised to companies and workers in eastern newspapers. Libbey had been following the great glass industry diaspora for several years. Personally, Libbey loved the society of Boston and its cosmopolitan environment. He was single and interested in finding a wife, and Boston seemed ideal for a young gentleman. Moving to Ohio seemed like moving to hell itself.

Northwest Ohio of 1880 was an area that America's manifest destiny seemed to have left behind. The area was dominated by the remnants of an ancient glacial lake known as the Black Swamp. In the first part of the 1800s, settlers drained the swamp for farming and transportation. The rich soil of the drained swamp brought German settlers to the area in the middle of the nineteenth century. The hardwood that thrived in the swamp created a lumber industry. Still, it was hard to see how Toledo real estate pioneer Jesup W. Scott (the grandfather of Libbey's future wife) made a success of a small book on Toledo titled *The Future Great City of the World* in 1868. Jesup Scott had once gotten lost in the swamp of today's downtown Toledo. Yet a canal system and Toledo's port offered potential equal to that of today's Chicago. Jesup Scott was co-owner and editor of *The Toledo Blade,* and he used that position to promote the area daily. However, Toledo had little to boast of except for its waterway connections.

The Ohio canal system was connected to Toledo in the late 1830s. It tied together the Great Lakes, the Ohio River-Mississippi River, and the Erie Canal in the east. The canal dropped shipping prices from $100 a ton to $8 a ton. The combination of the canals and Toledo's lake port opened the city up as a grain transportation center for Midwest wheat and corn, shipped to the big eastern cities via Lake Erie. By the 1870s, Toledo was the nation's grain center, ahead of Chicago. Toledo had even become the world capital of clover seed production. But just as Jesup Scott's dream was to be realized, railroads enabled straight shipping from Chicago to the East Coast. To gain business, the railroads made shipping rates cheaper from Chicago and Minneapolis to the East than from Toledo to the East. Toledo remained an important port, but its big city dreams declined. While passing through Toledo, the railroads helped the nation pass it by. The horse-driven, wagon-building business of Toledo declined with the advent of the railroads, and the brewing industry declined with the grain trade. In the late 1870s, Toledo and Northwest Ohio were again in search of their economic identity. Toledo had capitalized on its lake port to develop a major iron industry, but steel, which required coal, became the preferred structural material over iron. Pittsburgh soon eclipsed Toledo in the world of structural metal.

Toledo and Northwest Ohio citizens continued to vote for taxes to expand Toledo as a railroad center. Throughout the 1870s, key hook-ups were made with the eastern railroads. The Panic of 1873 dealt several blows to Northwest Ohio's rail expansion, but the area moved forward. By the late 1870s, grain was moving to Pennsylvania and coal to Toledo. Toledo politicians started to search for an industry to build around the railroad center. The area did attract some diversified industries after the Civil War. The combination of Germans and grain made brewing a major industry, and Toledo, at a quarter-million barrels a year, was not far off the production of St. Louis and Milwaukee. In the 1870s, Toledo had become a major wagon manufacturer, but by 1880, hardwood was rapidly being depleted in Ohio and Michigan, driving costs up. The businessmen and bankers of the area continued an aggressive program of attracting industries. Jesup Scott never gave up on his dream for Toledo. He teamed up with his sons (William, Frank, and Maurice) to promote building a university dedicated to the practical arts and workingman (Toledo University of Arts and Trades) in the 1870s. The school would be the backbone of future industrial development and would be a model for Andrew Carnegie's technical college (Carnegie Institute of

Technology) to support the steel industry in Pittsburgh. The Scott family had taken big steps toward the industrialization, but it took an unusual event to determine the future.

That event was the discovery of natural gas fifty miles to the south in Findlay, Ohio. For many years, oily films and marsh gas illumination was known to the locals. In April of 1884, Dr. Charles Osterlen, a local geologist, discovered gas in a test well on his farm. By December of 1884, Dr. Osterlen formed a natural gas company. The discovery spurred a drilling boom that resulted in a major gas strike in downtown Findlay on January 20, 1886. This Karg well produced 12 million cubic feet of gas daily on Liberty Avenue. The roar was deafening, and the flame could be seen at night in Toledo. With so much natural gas, streetlights ran 24 hours a day. Gas was supplied to stores and factories at $3 a month. In 1888, the Findlay Chamber of Commerce ran a national advertisement offering free gas to any factory in Findlay. Within a year, Findlay had 72 factories, 11 of which were glasshouses. Within two years, Findlay would have 31 glasshouses! So many factories located in Findlay that a critical shortage of workers resulted. Orphans were recruited from elsewhere for labor.

The great gas boom was underway in Northwest Ohio. Geologists suggested the gas pocket covered all of Northwest Ohio, and the geologic theory of the time was that gas was constantly re-generated. While test wells in Toledo had not been very productive, hope was abundant. The businessmen of Toledo were extremely active in the development of the city. Toledo needed natural gas to spur the economic boom that was happening just to the south, and even if Toledo lacked wells, it could still run a pipeline to Findlay. Charles Augustus King, one of Toledo's most successful businessmen, would lead the drive to bring gas to Toledo. King had formed the board of trade in Toledo and became the first president of the Produce Exchange in the 1850s. He was president of a local insurance company as well. In 1867, he was elected as mayor with the newly formed Whig Party and later was a state senator for the Republican Party. In the early 1880s, Charles A. King had formed a group to improve the lake channel and make Toledo a major lake port. In November 1885, he formed the Citizens Natural Gas Company with his son-in-law, David Locke, and several businessmen — Dennis Coughlin, M. Enright, P. Dowling, C. Lorenz, J. Kumler, and J. Hamilton.

David Locke had formed an influential business club in 1882 known as the Draconian Club (the forerunner of the present Toledo Club). He was the editor of Scott's *Toledo Blade* and was in a position to promote the area. Locke had written the *Nasby Letters*, a favorite book of President Lincoln. David Locke became the first president of the Draconian Club, and the first meetings took place in *The Toledo Blade's* offices. The hot topic at meetings was the failure of Toledo to attract new industries. The Draconians had strong political ties as former Whigs now with the Republican Party.

Another key member of the Draconians was George E. Pomeroy, an aristocratic banker and real estate mogul with business ties around the world. The formation of Citizens Gas and its ties to the Draconian Club sent chills through Pittsburgh capitalists, who feared the rise of gas use in Toledo would end their profitable coal trade to the city. A letter was printed in the Toledo paper from Pittsburgh saying: "The destruction of the Pittsburgh coal trade is indicated now by unmistakable signs. Unless the promised gas supply should unexpectedly fall, the next decade may witness the utter decay of the coal trade."[2]

Toledo's coal trade, rail system, and lake port gave it natural advantages. Still, Toledo had failed time and time again to recruit new industries. The Toledo businessmen were a determined bunch. Even before gas was discovered in Findlay, Toledo was courting glass companies. Glass companies would bring highly-paid, skilled labor and create many more jobs for companies supplying the glasshouse. A *Toledo Blade* reporter in 1884 estimated that a 200- to 300-man glasshouse would increase the Toledo population by one thousand.[3] In April of 1884, the Dithridge Glass Company was looking to move from New Brighton, Pennsylvania (down river from Pittsburgh), to Toledo. The failure to close the deal was blamed on a lack of capital because the city's funds were tied up in grain dealing. Many glass companies were looking for large incentives. Free natural gas or cheap natural gas was a real draw for glassmakers who were bound by their fuel costs. The muddy and swampy streets didn't help to attract capitalists, and Toledo businesses and citizens were putting a lot of money into urban development.

Unfortunately, the wells and test wells of the Citizens Gas Company failed to live up to expectations. The company then proposed a pipeline to the gas fields of Findlay, but lacked the investors, as fears rose about the amount of gas available. These Toledo businessmen had, however, created interest on a national level for the future of natural gas in the area. Cleveland and eastern capitalists, with the goal of building a pipeline from Findlay, formed Northwestern Ohio Natural Gas Company. Shareholders in Citizens Gas sold their rights to Northwestern Ohio Natural Gas. A second company, Toledo Natural Gas Company, formed by Toledo businessmen and financed by Pennsylvania oilmen, entered the race to build a pipeline. In addition, Cleveland oilmen were laying pipelines and building railroad tankers to supply oil to Toledo.

As the construction race of the gas pipelines continued through 1887, a new business organization emerged to support the industrial growth of Toledo. This group included Draconians David Locke, J. K. Hamilton, George E. Pomeroy, and Charles King, but a new leader emerged in William H. Maher, who a made a fortune in the hardware business. William H. Maher was a co-founder of the Toledo Natural Gas Company. Another important group was the banking community and the Ketcham family, which joined the effort of the Business Men's Association. The organization became known as the Business Men's Committee. It was part chamber of commerce and part real estate company, with a mission of bringing industry to Toledo.

Cities such as Bowling Green, Tiffin, North Baltimore, and Fostoria were reaping the benefits of natural gas by 1886. These cities had found success in attracting glass manufacturers to Northwest Ohio. Cheap fuel, rail transportation and plenty of land for industrial growth were key. These towns also had plenty of land; they were willing to trade for population and industrial growth. In 1880, North Baltimore, between Findlay and Bowling Green, was a mere junction station of the Baltimore & Ohio Railroad and the Toledo, Columbus & Southern Railroad. Local businessmen sold shares in a gas well and struck gas. The village fathers offered free gas and land. Within eighteen months, the town had four glassworks employing more than 500 men and boys. Findlay's population went from 4,500 in 1884 to 25,000 in 1889, as 31 glass companies moved to the city. Glass companies proved highly mobile. Buildings were wooden, and the equipment, including the furnaces, could be loaded on train cars readily. The skilled workers could be loaded on trains as well. In fact, many companies made two to four moves in the last quarter of the nineteenth century, following cheap energy west from the East Coast to Indiana. Toledo, however, seemed to be favored by only a few glassmakers.

Besides cheap fuel and transportation there were some other, lesser-publicized advantages for glassmakers. Many glassmakers, including Libbey, were aware of the Toledo sand deposits. Glassmakers had been purchasing sand from the Monclova pits since the Civil War. These pits shipped 100 tons of high-quality sand a day to the industry. Ohio capitalists had actually purchased the Toledo Glass Sand Railroad in October of 1887 in hopes of luring glass companies to Toledo. Test samples sent to Libbey in New England proved of high enough quality for flint glass. The raw materials such as sand made up only about 20 percent of manufacturing costs; thus, sand proved not to be a great enticement to glass companies.[4] This Toledo sand was a granular quartzite of good purity that could be used in window and flint glass. A *Toledo Blade* article suggested an annual savings as high as $10,000 a furnace pot, but this seems a bit exaggerated. While sand was an important factor to most glassmakers, Libbey felt strongly about the continued use of Berkshire sand, even if he moved out of New England. The Toledo sand deposit was more important to window manufacturers coming to Northwest Ohio. Glass required were soda ash, which was imported from England. Lime was also needed, which was abundant, high quality, and cheap in Northwest Ohio. Libbey, like most New England glassmakers, had to look west for the future, as the cheap energy was moving west. The market was moving west as well.

The westward rush of glass companies in the 1880s was also the result of favorable national policy. The 1880s had seen Congress's Ways and Means Committee dominated by three powerful, protectionist Republicans—"Pig Iron" Kelly of Pennsylvania,

James Garfield of Ohio, and William McKinley of Ohio. The glass companies were heavily protected, with tariffs as high as 45 percent. Ohio was supplying the votes and financial backing for these Napoleons of protectionism. In addition, the westward settlement movement was creating a huge market for tableware and lamps, as new households were being created. The industrial revolution was bringing in a flood of immigrants, who were starting families and building households needing tableware. Republican tariffs had created a demand for iron, steel, lumber, and glass. Department stores generated new demand for window glass. An emerging food industry needed glass jars and bottles.

Chapter 5

The Business Men's Association

Toledo was focused, and determined to get its share of the glass industry boom. The Toledo Business Men's Association (later the Citizens Committee of Toledo) had some advantages over the competing towns in the strong business background of its members. The Business Men's Association remains to this day a classic model for promoting civic and area development. Most of the members stood to prosper from the growth of Toledo. Men like the Ketcham brothers (Valentine,[1] John, and George), the Scott brothers, and George Pomeroy stood to make a fortune in local real estate. Bankers such as John and George Ketcham hoped to see a major influx of money. Hardware and lumbermen would reap the benefits of growth. But the entire business community stood to gain from new manufacturing companies.

The majority of members were merchants and small business owners. The Business Men's Association combined the functions of a Chamber of Commerce, local politicians, the Draconians, and other business groups with a single mission of economic development. The pillars of the Business Men's Association were real estate people, service industry people, and merchants, all of whom stood to gain most from growth. Glass companies, in particular, would bring well-paid, skilled craftsmen to add to the economic development. The association had allies in the railroad companies, who would see a substantial increase in business. The association represented the only true city in Northwest Ohio. The key officers were insurance agent and president of the association, John S. Kountz, and secretary William Maher. Three other key association members were money men—John Ketcham, George Ketcham, and George E. Pomeroy. Businessmen such as Toledo's "coffee king," Alvin Mansfield Woolson, and Toledo's "wheat king," Sheldon C. Reynolds, were also part of the association. In all, it had over one hundred members.

The Toledo Business Men's Association was a confluence of business, politics and culture. Economic growth was prized. The city had no Chamber of Commerce, so other organizations and associations had to foster economic development. The Produce Exchange had taken on this role in the 1870s. However, the Produce Exchange was focused on grain storage and shipping. The problem was that by 1880, Toledo was losing ground in grain management to Chicago and Milwaukee. The old political organizations and the banking system were closely tied to grain management. The merchants were seeing a decline and real estate had leveled off. The Toledo Business Men's Association included many merchants and real estate agents who had a different vision for the city. Generally, they were younger and more ambitious than their agricultural counterparts. They had captured the support of *The Toledo Blade* and many of the young social clubs. They represented the struggle between new and old. The bankers were starting to take notice of the trend toward industrialization in Northwest Ohio, but had strong, old ties to the Produce Exchange. Furthermore, the cities of Northwest Ohio had become locked in a competitive frenzy to attract new industries. Glass companies played cities against each other to obtain favorable deals.

The Business Men's Association smoothed the way to bring manufacturing to Toledo. It published flyers, advertised around the world, and sent members to speak around the nation. Politically, the association was strongly protectionist and Republican. The Republican Party at the time was strongly protectionist. The Produce Exchange, on the other hand, was Democratic and free trading. Manufacturers were protectionists, and farmers, railroads, and grain operators were free traders. This divided the city, but the association fought to show the benefit of manufacturing to all in the city. Hardware and real estate king William Maher became a leader in the Business Men's Association. He often published pro-Republican and protectionist articles in *The Toledo Blade*. Maher worked on an association committee to bring the state Republican convention to Toledo, and with a number of foreign manufacturing firms on potential moves. An outstanding writer, Maher wrote almost weekly articles loaded with statistics to support the need for manufacturing growth. In one, he argued that the 40,000 boxes of window glass used yearly by Toledo builders should be manufactured to everyone's advantage in Toledo.[2] Maher promoted Toledo in national magazines such as *Harper's Weekly*.

The Business Men's Association had created a real estate boom before any new industry set foot in Toledo. The *Chicago Tribune* reported, "Believers in Toledo real estate say that no mistake can be made in buying between Cherry and Monroe [streets], the river and Collingwood Avenue."[3] Most believed that with the gas pipelines rushing toward the city, an economic surge was impending. The real estate men created a boom psychology that creates economic bubbles. But there were some major roadblocks to industrialization. Most of the surrounding towns had direct or indirect control over the gas companies. Toledo had to depend on piped-in gas. The two pipeline companies in Toledo were private, and their majority investors weren't from Toledo. The pipeline companies expected to make a profit from the sale of natural gas and supplying gas to homes was more profitable than supplying industrial customers. Initially, companies planned to sell gas slightly below the cost of Pittsburgh natural gas and at 75 percent the cost of coal, but the deal varied for each

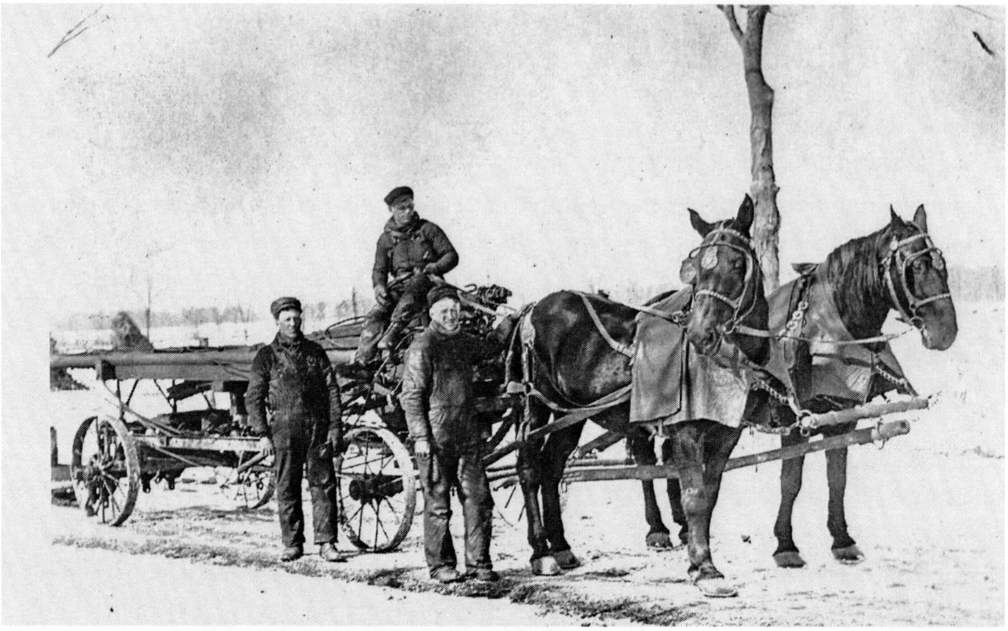

Laying of gas line from Findlay to Toledo in 1887 (courtesy of University of Toledo, Ward Canaday Center, Libbey Glass Records).

company. The association lobbied with the gas companies over rates for the industrial prospects.

Members of the association realized early on that finance was going to be a major problem. The Produce Exchange had most of Toledo's capital and banks invested in grain exchange expansion. Eventually, association members formed Union Savings Bank. Its first board of directors meeting was held in the offices of *The Toledo Blade* in 1888. James Secor was elected president and William Maher was elected vice president with association members, it was clear that the Business Men's Association was a major force in the economic development of Toledo. New England Glass was considered a major prize of the campaign.

The first contact appears to have been initiated by Edward Libbey, who was actively searching for a new location by the summer of 1887. Even prior, Edward had been exploring new locations for his father. Toledo was making headlines at the time by hosting the Ohio Republican Convention. Libbey was answering the advertisements of Northwest Ohio towns, which were flooding the eastern newspapers in 1887. Libbey had contacted Tiffin, Findlay, and Fostoria prior to responding to the Business Men's Association. He visited Tiffin in the early fall of 1887. It offered $25,000 cash, free gas and land. Libbey felt Tiffin lacked the cultural and social resources required. At the time of Libbey's visit, Tiffin had no glass companies yet, but Libbey was aware that several Pittsburgh and West Virginia glasshouses were hoping to locate there.

The nationally known A.J. Beatty & Sons was also considering Tiffin and would move there the next summer (1888). Libbey proved to be a very cautious businessman, believing the hype of the local papers about endless gas supplies. So he viewed the Tiffin visit as a start of a process of detailed evaluation. Libbey was also very interested in Pittsburgh, which offered a city environment more like that of Boston.

Pittsburgh and West Virginia were bidding for glass companies. Pittsburgh, in particular, offered many advantages for New England glass makers. With an abundance and variety of fuels, being fifty miles from the world's greatest known coalfields and oilfields. In addition, it had an abundant supply of natural gas and the world's best pipeline systems. Pittsburgh had banking, transportation, and capital advantages as well. In 1886, the city of Pittsburgh was the glass capital of the world. It had 50 glasshouses and produced 90 percent of the nation's window glass. Its annual glass show at the Monongahela House was the largest in the nation. It employed over 5,000 skilled glassworkers, including the young sons and wives of steelworkers. It had a raw material supply network in place, and a shipping distribution network of railroads and rivers. But Pittsburgh refused to get into bidding wars with the Ohio and West Virginia towns. The towns even offered cash along with free land and fuel. Free gas was worth about $3,000 to $4,000 a year. Gas companies in Pittsburgh were in business, not in community promotion. Coal in Pittsburgh cost about the same as natural gas in Pittsburgh. Ohio towns were giving away cash, and amounts in the range of $5,000 (over $100,000 in today's dollars) became commonplace. The lure of free land, money, and fuel was pulling all kinds of investors. This "manufacturing rush" was not sustainable.

The Toledo Business Men's Association's advertisements in 1887 had brought a flurry of proposals, including from three large glasshouses—Libbey's New England flint glass works, Charles Hurrle's window glass company, and a glass bathtub manufacturer. Like the gold rush, the glass boom of Ohio, Indiana, and West Virginia was bringing a lot of characters with get-rich-quick schemes. Pittsburgh investors were quickly forming new companies to take advantage of free land and fuel to the west. The Toledo Business Men's Association had to deal with many of these characters.

About the same time the association was dealing with Libbey, a wealthy southern gentleman, "Colonel" Christopher McLean, appeared. The frenzy caused even conservative businessmen to be pulled into less-than-sound proposals. In the case of the "colonel," he had no real manufacturing experience but played up his knowledge of chemistry and glass. He was able to gain the backing of George Ketcham, the son of one of Toledo's founding fathers. George Ketcham was a city councilman, gas pipeline investor, and banker. McLean had big ideas, proposing the manufacture of glass bathtubs. He was able to get an advance to purchase a 94-acre lot, which he quickly subdivided into 500 lots for sale at $200 each,[4] to raise money for the factory. Sales were minimal in the conservative town.[5] McLean eventually built a factory that ran a mere three months before closing.

Clearly, with the McLean deal, the association was caught up in the rush to land a glass company. The hesitant real estate community and many bankers demanded that the Business Men's Association take a closer look at these prospects. Charles Hurrle's proposal was in the works before Libbey ever visited Toledo. Hurrle was a seasoned manufacturer of window glass, which represented the largest growth sector for glass at the time. The building boom and the rise of department stores had created demand. Window glass manufacture at the time was a high-skill and labor-intensive specialty. Shipping and packaging costs were significant, making location near the market a key factor. Pittsburgh was the center of the country's window glass industry. Shipping cost by rail to Chicago was 14 cents per hundred weight compared to 10 cents per hundred weight from Toledo, or about a 30 percent savings. A typical 10-pot furnace in Pittsburgh would cost $5,000 in natural gas, compared to $3,500 in Toledo. The best quality window glass was imported from Belgium and France. Because window glass was considered an infant industry needing protection, William McKinley, chairman of the Ways and Means Committee, was able to get better tariff protection in the early 1880s. In 1887, window glass was the most profitable segment of the glass industry. Toledo bankers seemed more than willing to back Hurrle's factories. He got a mere $6,000 and an acre of land.[6] For Hurrle, Toledo's freight, sand, and fuel advantages were the tipping point. Still, the Business Men's Association had to sell the town aggressively. A few months earlier, they had lost a similar window operation to Bowling Green (15 miles to the south), which offered $4,000, free land, and free gas. It should be noted that glass companies moving from the East Coast could expect a 10 percent increase in wages in the west.

Libbey had traveled to Europe in May of 1887 to sell the patent rights for his Amberina. He attended Queen Victoria's jubilee celebration, and spent the summer touring museums and glasshouses. He spent time observing cut glass technology, and visited England, Scotland, Italy, France, Germany, Ireland, and Switzerland. Europe was the trendsetter, and the Great Exhibition of London in 1851 had centered on glass. The Crystal Palace was the greatest glass building ever built. More importantly, this world fair launched the rage for fancy glass tableware. While the trend was centered in England, many of the techniques were from the glasshouses of Germany's Bohemian forests, which Libbey visited. At Vienna and Antwerp, he spent time studying art and paintings. Here, for the first time, Libbey's love of the old masters surfaces. Libbey made notes on the beautiful architecture as well.

The trend in European glass had changed from its emphasis on cut glass. Enameling, staining, decorating, and coloring of glass had evolved out of the same old Bohemian factories. Libbey visited old friends at Stourbridge, England, and the famous crystal factories of Murano, Italy. There he studied *cristallo*, a very pure soda-lime glass, as well as *latticinio*, a very special mixture of milk and clear glass. The famous English glassmakers were, in fact, Bohemian artists. The wages of these outstanding Bohemian craftsmen were a quarter what they could earn in the United States. Libbey realized that with the major cost in art glass being wages, the Europeans had a huge advantage. Staining created glass pieces with an array of colored scenes, cameos, and landscapes. Colors such as milky opal and marble were popular in the European pressed glass market. One of the new trends Libbey observed was iridescent glass. Staining with metallic oxides produced iridescence. Libbey loved the art and color of European glass but realized its market would be limited in the United States, which demanded more utilitarian characteristics.

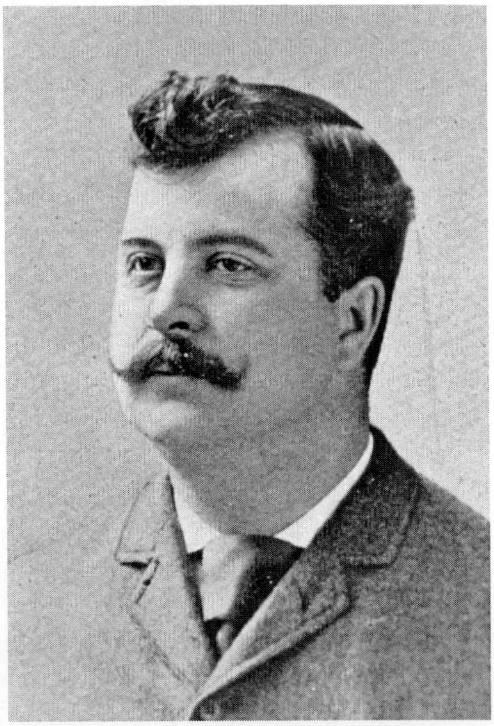

Edward D. Libbey, 1880 (courtesy of University of Toledo, Ward Canaday Center, Libbey Glass Records).

The trip probably honed Libbey the collector more than Libbey the glassmaker. He saw more automation in the production of this art glass, not on an industrial scale but at the crafts level. Simple lathes and tooling aided the artist. He visited many museums and noted paintings in his travel diary. He visited cathedrals in Antwerp, Milan, and Cologne. His notes show the passion that would, years later, bring him to found a museum. Like his father, Libbey loved to travel. He renewed acquaintance with old friends such as A.J. Nash in England, and gained many new ones. Unfortunately, in September, Libbey had to return to East Cambridge to face the financial difficulties of the business.

When Libbey first saw Toledo that fall, it had a population of around 57,000, compared to Boston's 420,000. It had many advantages, but the Business Men's Association was losing out to the towns of Findlay and Bowling Green, which could offer free natural gas and cash bonuses. Free gas could mean a savings of $5,000 to $20,000 a year. Still, the boom psychology made it a glass company market, and they were demanding free land and large cash payments in addition to free fuel. Libbey spent most of the fall exploring possibilities in Pittsburgh, West Virginia, Ohio, and Indiana. He traveled with the company's secretary, Solon O. Richardson, and at times with employee and friend Jefferson Robinson. These men were conservative and knew tough times. Libbey's view of the art glass business was a bit different. He was leery of the natural gas promises, and the ability of these towns to deliver, long-term. Furthermore, Libbey needed craftsmen to make his cut glass, and these craftsmen wanted the lifestyle of the larger cities.

He also needed a large number of skilled workers: glassblowers, gatherers, mold makers, cutters, etchers, and engravers. He believed he needed to move a core of nearly 50 skilled men with equipment for a total workforce of 300. He would have to entice them and pay for the move. In the initial trips to Ohio, Libbey brought some of his key craftsmen. He realized that small towns like Tiffin and Bowling Green would not meet their needs. Many of these craftsmen were Irish Catholic and Toledo had an established Irish community going back to the canal building days. There were two large Catholic churches and even Catholic schools. Most of the smaller towns tended to be German Lutheran. Toledo also had a solid public school system, while many towns like North Baltimore and Tiffin had one-room schoolhouses and no high schools. In addition to the skilled craftsmen, Libbey would need to hire 60 to 80 lower-paid boys and women. Towns like Bowling Green and Findlay already had few available boys.

Libbey observed another important factor in favor of Toledo. It had a well-organized, professional fire department, and had invested over a million dollars in a pumping and water system to support fire fighting. The concern was paramount, as a November 1887 fire had burned the Hirsch-Ely Window glass factory to the ground in Findlay. Fires were common in the industry, with its high temperature furnaces and wooden structures. Gas-fired glasshouses were particularly prone to fires. Towns like Findlay and Bowling Green lacked fire fighting resources. Insurance rates were increasing, and loss of production often crippled companies after factory fires. Fires were feared as much by the workers, who lost months of wages. The concern would prove justified, as over twenty Ohio glasshouses would burn by 1892.

Libbey also realized that many men would refuse to leave Boston without the enticements of pay increases and help with the move. One of his demands was free lots for the workers on which to build homes. He needed a city that offered the recreational and cultural activities demanded by skilled craftsmen. With its manual training university and larger population, Toledo did offer the prospect of future glassworkers. But Pittsburgh was still in the running. The decision came slowly, with Libbey making many trips to Toledo. J. Kountz represented the Business Men's Association, with Secretary William Maher and George E. Pomeroy doing most of the work. Kountz headed up a sub-committee for contributions to buy land for the project. Negotiations were one-on-one and then with the association's full membership and important politicians. Many of the bankers and businessmen were hesitant, preferring to see Toledo remain a grain center. They did realize that New England Glass had a world-class reputation and more potential than most of their other prospects.

One of Libbey's November trips was to finalize factory sites. George E. Pomeroy, a Draconian Club member and association member, was the Toledo real estate representative assigned to help Libbey. Pomeroy had extensive business ties across the nation and was a well-known art collector, something that would later be of interest to Libbey. Pomeroy also had strong ties with Toledo's real barons, the Scott and Ketcham families. Finding a site was hard. Toledo consisted of swampland, canals, and wooden locks. Mud was everywhere. There were few paved roads and most were wooden plank roads. What good roads there were had to be lined with drainage ditches. Malaria, or "swamp fever" was common. Cholera was always a fear with poor drainage and sewage systems. Pomeroy was able to satisfy Libbey with a piece of land on a natural knoll, which was well drained. The knoll was at the intersection of unpaved Ash and Buckeye Streets, overlooking the old canal. The land was crossed by the Wheeling and Lake Erie railroad, which had direct access to the new Ohio coalfields of the east. This direct connection to the Ohio coalfields allowed Toledo to have quality coal at a lower shipped price than it would cost in Pittsburgh. The four-acre lot was being used as a baseball field. The land was valued at $4,000 and had nearby lots for the workers to build homes. The site would be supplied with natural gas by the Northwestern Ohio pipeline. Pomeroy was active in negotiating a lower price for Libbey.

The price of natural gas in Toledo was becoming a major problem. M. J. Enright, the president of the Business Men's Association, had again suggested a municipal pipeline to reduce costs to manufacturers. Enright lost a battle with the state to offer a bond issue but continued to push for a third company to supply gas to manufacturers at nominal rates. Enright envisioned a cooperative company that would supply gas without trying to generate a profit. However, Enright and the association failed to raise enough capital. Still, Libbey was not as concerned as others about the gas prices. He was paying $12,000 a year for coal in New England, and a Toledo gas bill of $3,000 didn't seem like a problem. Also, Libbey had assurances that gas would continue to be delivered below Pittsburgh prices. He valued the other advantages of Toledo over free or cheap gas and was happy that the gas price was lower than that of Pittsburgh. For his part, Pomeroy got Northwestern Ohio Pipeline to make a $500 contribution to purchase the four-acre site for $3,488. Northwestern also would supply the related worker lots known as Glassboro.

By December of 1887, negotiations were being finalized. Maher had traveled to New England to see the factory in operation and settle some details. Libbey was probably down to three choices — Toledo, Pittsburgh, or remaining in Boston.[7] Maher, Libbey, and their representatives made a number of back and forth trips throughout December and early January. Details were presented each time to the full association meeting, which slowed the process. For his part, Libbey had agonized over every detail and seemed to be hoping for some type of miracle. He realized the decision would have to be made soon, since his Cambridge factory's lease would be up. On February 6, he would have to notify the leaseholder of his plans to leave. The American Flints kept the pressure on about wages, but fuel was still the bigger issue. Still, Libbey could not take a wage increase at the East Cambridge factory. The national union, controlled by western producers wanted wages up. Libbey patiently negotiated with the union and used his foreman to help him make the case with the local workers. But in reality, even a major wage reduction would barely make him competitive. Another union issue was work rules which would limit production and hours worked. Libbey had countered with some different crew arrangements that could reduce costs. These final negotiations broke down, and Libbey realized he had to leave New England. Meanwhile the union called a strike and about forty men joined in. Libbey's response was to "give them but one week in which to return, to come back at the old rates, after which all who do not come in will be refused work again at our works."[8] The strike would be the final straw, and Libbey made the decision to go to Toledo.

Chapter 6

Florence Scott

Libbey's hesitation was as much personal as business-related. He loved the social scene in Boston, as did his sisters, who were all living with their stepmother in the Libbey house in Boston after his father's death. Libbey also enjoyed the cultural opportunities available in Boston. His older sister Alice had married Boston hardware and furniture magnate, William S. Walbridge; but Sarah would be moving with Libbey. Sarah had seen Toledo on one of Libbey's visits, and while not impressed, she wanted him to succeed in the glass business. Sarah had seen her father struggle with New England production costs. Libbey had tried to get his father to consider moving west.

Of all the stories surrounding the move of New England Glass to Toledo, the story of Libbey's attraction to Toledo's Florence Scott is underestimated. Florence may well have eased Libbey's concerns about leaving the social scene of Boston. Libbey had been a member of Boston's elite Calumet Social Club. But Florence could point to the new Toledo Club, which included her father, George Pomeroy, and the Ketchams; as well as the West End's Toledo Outing Club. Florence had a passion for collecting art and old glass pieces interested both Libbey and her. Florence was European-educated and as well-traveled as any Boston girl. In December, Libbey seemed to be searching for reasons to select Toledo over Pittsburgh. There is no question that even with the failed union negotiations, Libbey hesitated to make the final decision. The Business Men's Association, however, was determined to bring Libbey to Toledo. At 34 years of age, Libbey was struggling to become the CEO and leader demanded of him. While deliberate, he proved to have the skills of decision-making and broad analysis required for a major strategic change. The agreement would finally be signed in February of 1888.

In retrospect, Florence Scott might have been the determining factor. Libbey was 32 when he met her; and in Victorian society, he needed to take a wife. During his visits to Toledo, he met Florence in the office of George Pomeroy. She was the daughter of Maurice Scott, the son of one of Toledo's wealthiest and powerful families. Maurice Scott was a major real estate owner in downtown Toledo. Her grandfather was Jesup Scott, one of Toledo's greatest supporters, and co-owner of *The Toledo Blade*. Jesup also had founded the Toledo University of Arts and Trades and was a major real estate holder. Florence had lived in New York most of her life and had extensive family ties in New England. Her father had been born in Connecticut and her mother had been born in New Hampshire. Her stepmother was from Boston. Florence had made many trips to Europe, and she had gone to finishing school in Heidelberg, Germany. While in Germany, Florence made friends with Rosa Cristine Lang, who would play a key role in the lives of the Libbeys.

Florence and Libbey met after as Libbey was deciding on his move to Toledo. Florence would be invited to have dinner with Libbey at the West End home (806 Huron Street) of George Pomeroy, and she helped highlight Toledo's cultural and social scene as well as Pomeroy's beautiful art. She was known for her striking beauty. Many years later a friend remembered: "The key thing about Florence Libbey was her great beauty. I can remember even her perfume. She would come to our house with her hair just so and her furs and her coat. She had a very lovely laugh. Her tact was terrific."[1]

During Libbey's November visit, Edwin Booth (Lincoln assassin's brother) was staring in a touring version of *Julius Caesar* at the Wheeler Opera House. The Wheeler Opera House was on the corner of Monroe and St. Clair streets and had been attracting national productions since 1871. Toledo also had professional horse racing, baseball, and yachting. While they were meager offerings compared to activities in Boston and Pittsburgh, they were far superior to what was available in towns such as Findlay and Tiffin. The Scotts were members of Toledo's First Congregational Church, which still had New England roots. Libbey joined the Congregational Church, which had strong ties to New England, but he was never extremely active. Libbey also joined the Masons, after belonging in New England. Toledo certainly offered more business and social opportunities than the surrounding Ohio towns.

The final deal was hammered out in January. Libbey was hesitating, not over a better deal, but concerns about the agreement. The deal required $4,000 (about $150,000 in today's dollars) to be collected from citizens of Toledo to purchase a four-acre lot for the factory and fifty lots for worker homes. Even when Libbey finalized the deal, the Business Men's Association was still more than $500 short. Libbey had an informal agreement with the gas company to supply gas at less than the cost at Pittsburgh. Maher had a strong selling pitch — that the donations would, in turn, increase Toledo real estate prices by 25–30 percent, which indeed happened.[2] Men like George Ketcham, who contributed $100 to the Business Men's Association, had already made significant real estate gains on his Toledo land. Furthermore, on the fifty donated lots, worker homes would be built for $2,000 to $6,000 ($41,000 to $123,000 today). In addition, these high-paid workers would spend freely in the city. Old banking families such as the Ketcham family were quickly becoming supporters of the glass industry. John Ketcham (1832–1897), the brother of Valentine and George Ketcham, had even opened a new bank (Ketcham National) in January, hoping to serve glassmakers. The gas company would sell gas to those worker homes. Over one hundred men would be hired to build the factory. The Hurrle Window Glass Company and the McLean factory would be located in the Glassboro area as well. Libbey would build a three-story brick glass glasshouse and transfer $25,000 worth of equipment from New England.

Libbey planned to build a substantial, fire-resistant factory, not the typical wooden structures of most of these newly transplanted glass companies. To raise the capital to move and build, Libbey issued 2,000 shares of stock at $50 a share. He kept full control by personally retaining 1,598 shares. Libbey borrowed from his sister Sarah, his brother-in-law William S. Walbridge, and his uncle's family. Once the stock and agreement were settled, the company could then go to local banks for more financing. When Edward Libbey and Solon O. Richardson came to sign the agreement, they brought Pittsburgh architect Charles Hamilton with them. Hamilton was from Wheeling, West Virginia, and had built twelve glasshouses previously. Libbey braved the damp cold of Toledo to go over the building details personally in February.

The Libbey factory would be the palace of Northwest Ohio factories. Its Edwardian brick architecture would rival that of the H.J. Heinz ketchup factory in Pittsburgh, considered the world's best factory building. The building would use iron, steel, and stone as a substructure, making it nearly fireproof. Iron doors and a corrugated iron roof substantially added to the fire-proofing. The fireproof design was written up by many engineering journals of the time.[3] The building was heated by steam and had two elevators. The core of the glasshouse was a 13-pot furnace with 3,000 pound pots, making it America's largest. The furnace would be state-of-the-art, capable of using coal, oil, and natural gas. Libbey intended to switch according to the going price of these fuels. Toledo's railroad connections assured carloads of coal at Pittsburgh prices and tank cars of oil at Cleveland prices. Delayed by the Great Blizzard of 1888, construction began on St. Patrick's Day. By mid–March, the new area of Glassboro boomed with construction of Libbey's factory, and Charles Hurrle's factory next door. The total cost of the Libbey building would be $50,000. Libbey put up the money himself, using family money. His board (which he controlled) authorized a payment of $80,000 to Libbey on April 14, with the cost of the building and equipment transferred from East Cambridge. Libbey was far from free of financial problems; he would need to invest for several months to make payroll and meet other costs.

Libbey had hoped to build a glassmaking "village" like that of Sandwich, Massachusetts, but money limited that dream. He was struggling to get workers to leave their beloved Boston, always a concern. His idea of a worker village, he hoped, would entice more skilled workers to follow him to Toledo. Libbey was highly dependent on his skilled cutters. Louis Vaupel (1824–1903) was his best engraver and had been with the company since 1850. Vaupel's work was considered museum quality, on a level with the masters of Europe. His Masterpiece Chalice at today's Toledo Museum is considered his best work, but his pieces are in fourteen of America's best art museums. Vaupel's sick wife and strong family ties made it impossible for him to come to Toledo. Libbey arranged for him to work in Boston as a commissioned employee.

Contrary to some accounts, the market for Libbey Glass was picking up in the 1880s. In fact, the 1880s were times of national economic expansion. A booming middle class of merchants and skilled workers was trying to emulate the spending of the wealthy. Many a lower class family might have a single piece of cut glass as a centerpiece for the home. These middle class families were reading magazines like *Good Housekeeping*, *Harper's Bazaar*, and *Ladies' Home Journal*, which created a style market in household goods. In addition, department stores and mail-order catalogs were evolving and creating a bigger market for pattern glass. In the 1880s, Libbey tapped into this market trend. Libbey's call bells used to call guests to dinner, were becoming common in middle class homes. Linen napkin rings of cut glass also became popular with the middle class. A table set usually included a sugar bowl, spoon holder, creamer, salt cellar, plates, napkin holders, goblets, wine glasses, and water glasses. Knife rests became a popular part of the table setting as well. The knife rests, bells, and rings would be in the same pattern as the tableware. Libbey continued to create new products to the very last day of the Cambridge operation. The sales catalog was used to show prices to the end customers. Libbey salesmen would sell to places like Wanamaker's Department Store in Philadelphia at half the listed price, allowing a 100 percent markup. Typical prices for cut flint at Wanamaker's in 1880 were $12 to $20 for a dozen tumblers or goblets. Pressed glass tumblers and goblets were $2 per dozen.

What Libbey and his managers Locke and Hopkins lacked in operational skill was more than made up in strategic business skills such as marketing, product development, and finance. While Libbey was making the transition and move to Toledo, he was working on a pattern and product line with Joseph Locke at the New England factory. It was the now-famous Maize art glass line. The 1880s represented a period of economic growth throughout the nation, creating a booming market in glass tableware. There was a rising middle class and even a worker class that had disposable income. Tableware was popular, and people tended to buy a pattern for their household. Pressed glass makers had successfully tapped into this market. Colored glass patterns were also a fad of the period. The Maize line of New England Glass was to be a partially mold blown and pressed line. The pattern would come in all forms of tableware including salt shakers. Libbey planned to go a step further in offering Maize lamps. Locke was developing this cornhusk pattern in early 1888. It was an opaque glass of ivory or light yellow. The husk part was stained with green and sometimes blue, red, or orange. The glass pieces were thin and aimed not at the middle class, but the lower pressed glass market. Staining allowed Locke to reduce costs, but it required some development work. Stain was brushed on and fired in a special furnace (a muffle furnace) at temperatures above 1,000 degrees. Libbey would launch a major advertising campaign after he got his new Toledo factory running, and the initial sales in the summer of 1888 were extremely encouraging.

The new operation had a simple management structure. Libbey was president and treasurer (although Bill Donovan was handling the details), Solon O. Richardson was secretary, and Joseph Locke was the operations manager. Jefferson D. Robinson was the chief sales agent, located in Chicago. Libbey planned to use a foreman management system more acceptable to the union in the West. By June, Libbey had started the final plans for the move to Toledo. He offered to take his workers with him, but many hesitated. In the end, over one hundred employees were willing to at least try the move. However, some key glass workers refused, forcing Libbey to advertise in Pittsburgh and West Virginia for skilled workers. The first contingent of glassmakers came in late June to start the furnaces and make trial batches. The glassmakers soon realized that the heat and humidity of Toledo would be worse than in New England. Libbey brought sand from New England and imported clay for pots. Joseph Locke and John Hopkins worked with the gangs to test the product. Pressed glass items and glass tubing were made early to proof the furnaces. It took some time to get used to the gas heating. Gas heat tended to be fast, hotter, and cheaper.

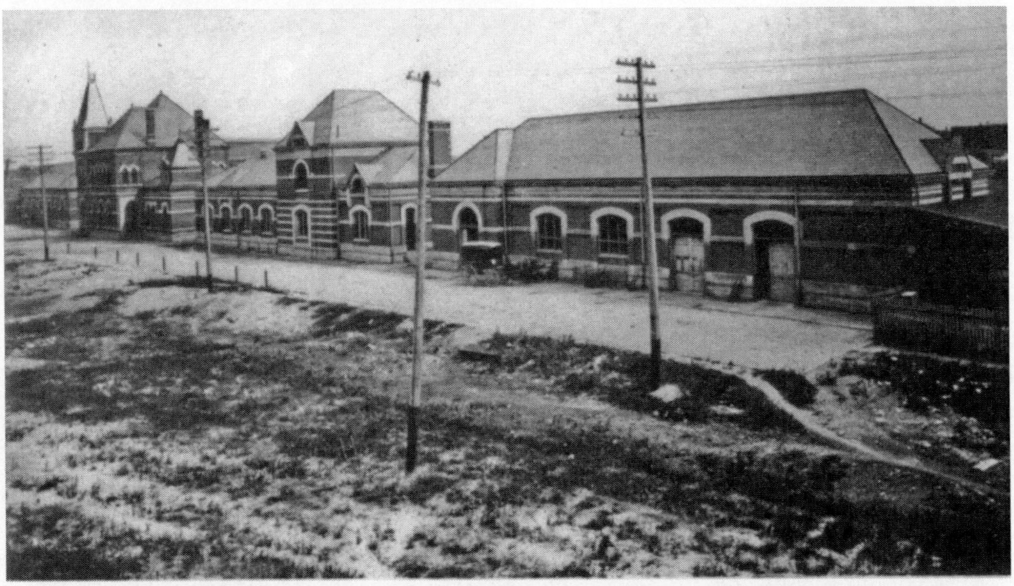

Toledo' s Union Station in 1888, where Libbey and his workers arrived on August 17 (courtesy of University of Toledo, Ward Canaday Center, Libbey Glass Records).

During July, hundreds of Toledo residents viewed the beautiful Libbey cut glass products at Bell & Powell's store on Monroe Street. Boarding houses and hotels were gearing up to offer temporary housing for the incoming workers. While sales of the lots were still not completed, some excitement was building for Libbey's arrival. On August 16, 1888, a little over one hundred Cambridge glassmakers boarded the train for the thirty-six hour trip to Toledo. Most of their families were left behind. The train would make several stops to pick up new glassworkers from West Virginia. A drizzle greeted the train at the Toledo station, where hundreds of Toledo citizens gathered to welcome it. City officials had prepared a major celebration, including a local military band and a long list of speakers. Mayor Hamilton, with association officers J. Kountz and W. Maher, joined him to extend their hands as Libbey stepped off the train. This was the finale of months of negotiating and compromising.

Solon O. Richardson and William Donovan, accompanied by Libbey and Jefferson D. Robinson, joined them from Chicago. The arriving workers then formed a parade to march through the city to the Ash Street factory. The company had prepared posters in anticipation of the parade. The streets were lined with happy spectators. In addition, about sixty unskilled boys and women from Toledo who had been hired joined in the parade. The paper reported hundreds had applied for the sixty positions. The parade ended at a wooden bandstand at the factory with many citizens joining in. Libbey took the stand and thanked the city for a great welcome. He also presented Mayor Hamilton with the first piece of glass made at the new glasshouse. After some typical political speeches, the crowd was served a picnic-style meal. Libbey was said to have worked the crowd with the politicians. He understood, as did W. Maher, that a factory and a community had a symbiotic relationship. Toledo was already experiencing a construction surge as a result of incoming glass companies. The growth of Toledo and Libbey Glass would be linked for many decades to come.

The factory was running the next day, but struggling was probably a better description. Libbey had market problems to contend with. He had lost $40,000 the year before, mainly due to labor, lost market share to pressed glass, and fuel issues. Pressed glass was rapidly eroding the Libbey cut-glass market. The bankruptcy of cut-glass maker Bakewell Glass of Pittsburgh had demonstrated how pressed glass was destroying the cut-glass market. Many of the remaining cut-glass makers were moving into pressed glass lines to stay profitable. As much as 85 percent of the blown glassware had been replaced by pressed glass. Pressed glass had even knocked out most of the china market. While Libbey was primarily a cut-glass manufacturer, the company did manufacture some pressed glass. Personally, Libbey wanted to stay in the prestigious cut-glass market. The economy in 1888 was very strong, with most industries making money. The glass industry, however, was over-expanding, even though market demand was growing. The over-expansion created a bubble, the seeds of the "tableware depression" of 1890 were sown, as over one hundred new glasshouses

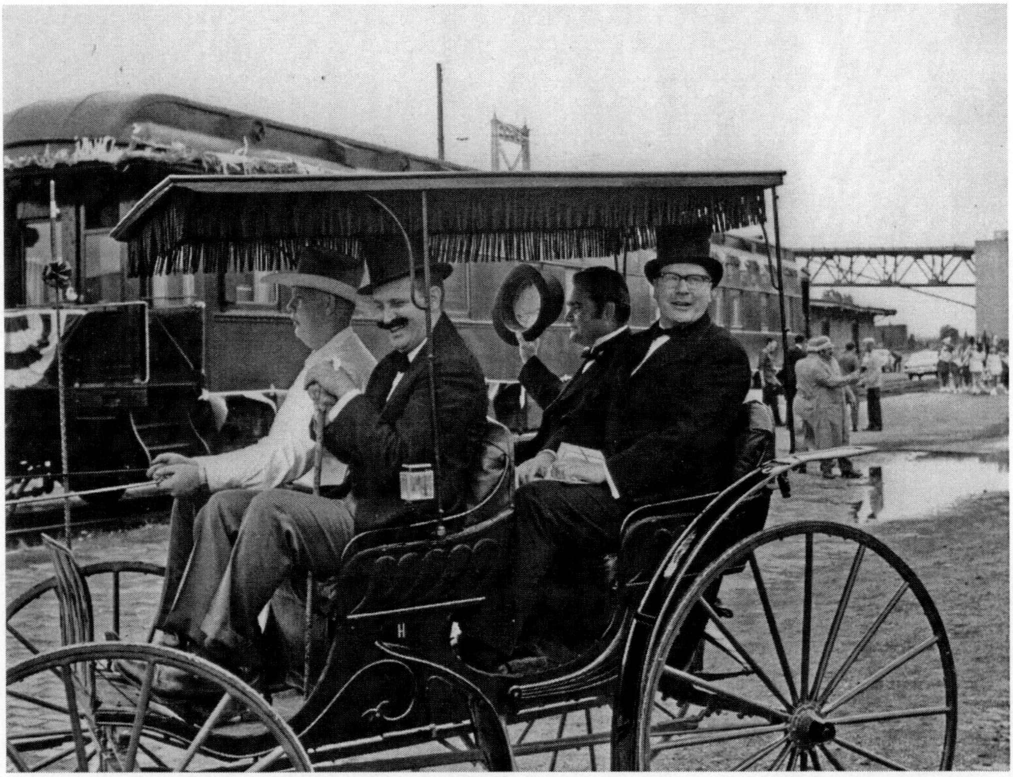

Reenactment of Libbey's arrival at Union Station in 1888 (courtesy of University of Toledo, Ward Canaday Center, Libbey Glass Records).

were constructed during the natural gas boom. Libbey had plenty of orders in 1888, but he was missing delivery due to poor quality at his new Toledo operation.

By the end of the August, Libbey had shipped very little, and production continued downward in September. The furnace could not be controlled properly, batches were being overheated, and in other batches, inconsistencies occurred. Libbey focused on lower quality pressed glass, tavern tumblers, electrical insulators, railroad lantern globes, and low-cost tableware. Less than 10 percent of the product could be shipped, and almost none of the flint, leaded cut glass. Things improved a bit in October, but union issues created new problems. Superintendent Locke and Libbey had accepted the authority of the American Flints. Some concession was necessary, since Libbey had around 30 to 40 West Virginian glassworkers. Some workers had returned home to Boston. The workers got the higher wage scale of the western, union employees and the shorter work schedule. The result was actually a lower take-home pay. Morale was destroyed. The men found more time on their hands and filled the taverns and saloons. Locke, while a great artist and designer, lacked the management and leadership skills to turn things around.

Libbey didn't know it at the time but he already had the solution in hand. In late August, to fill the jobs left by those who returned to Boston, Libbey had advertised and went to West Virginia to recruit new glassmakers. By the end of September, Libbey had lost a third of his Boston workers. One of the men Libbey hired was Michael Joseph Owens. At the age of twenty-nine, Owens was an experienced assistant glass blower. He had started at the age of ten as a furnace boy at the Wheeling glasshouse of Crystal Company, and then moved to Hobbs, Brockunier, and Company. He became a mold boy for the blowers and then moved up the crafts hierarchy to assistant blower. When Libbey interviewed him, Owens was working at several companies as a glassblower. He had hoped to be hired as a foreman by Libbey but had to settle for an assistant blower job ("servicer") on one of the factory lamp blowing gangs.

Owens had been active in the American Flints as a union officer, but he was ambitious and dreamed of management and ownership. As a union officer, he had shown his leadership skills. Owens was a forceful floor leader and wasn't afraid to use his fists. An Irishman himself, Owens could take charge of the tough Irish workers. This would prove a major advantage because Irish workers of the time had problems taking orders from English bosses. His operating experience included almost every possible position in a glassworks, particularly in natural gas furnaces. None could be better at training inexperienced mold boys and other boys. The only hands-on experience he lacked was in glass cutting. Owens's experience in gas-fired furnaces was something that Libbey needed badly. The furnace issues were at the heart of Libbey's operating problems.

The glasshouse that Mike Owens came to in Toledo was in disarray. Homesickness had devastated the gangs, and continuing furnace problems were demoralizing the remaining workers. Discipline was lacking, as workers drank hard to relieve their homesickness and work problems. Hung-over workers were late to get the batches melted, forcing gangs to sit around and wait. New boys who needed direction and training fell into bad work habits. The cutting and engraving departments were starved for work, and cash flow was a trickle. Managers like Locke and Hopkins were great artists but lacked leadership and managerial skills. The biggest problem was in the leaded flint cut glass product. As winter approached, many feared the company would not survive. The new West Virginian workers, however, were starting to get a handle on the furnaces, with their natural gas experience. But the West Virginians were not mixing well with the New England glassmakers.

As Libbey struggled, a serious competitor had opened 45 miles south in Findlay, Ohio. The company was Dalzell, Gilmore & Leighton Company, founded by Pittsburgh bankers, industrialists, and glass men. The Dalzell brothers, James and William, had a background in art glass and finance. E.D. Gilmore was an officer at the Pittsburgh Trust Company. These industrialists owned a very profitable West Virginia glasshouse, but the enticements in Ohio were too great to pass up. The new company

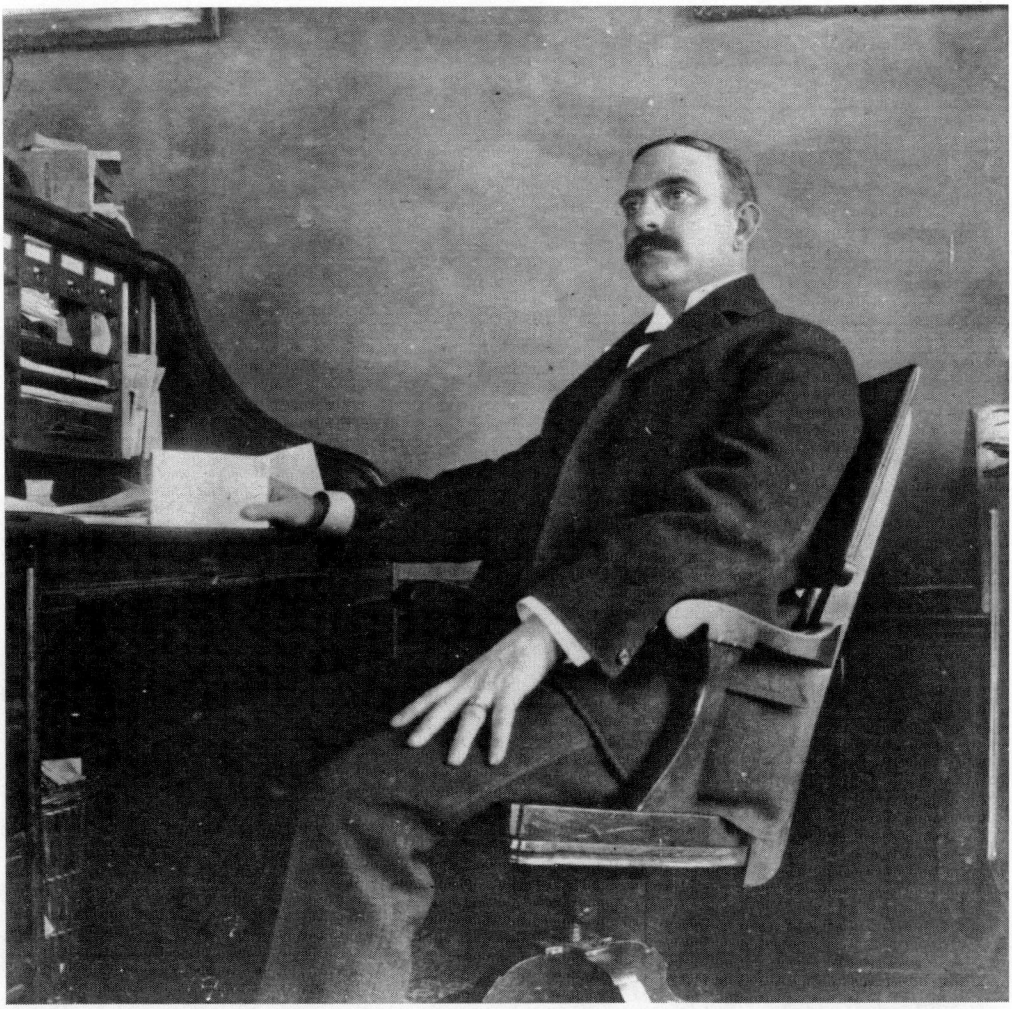

Jefferson D. Robinson, 1889 (courtesy of University of Toledo, Ward Canaday Center, Libbey Glass Records).

was offered free gas for five years, a three-acre lot for the factory, and $5,000 cash. Three members of the famous Leighton glass family — William Sr., William Jr., and George, representing father, son, and grandson — augmented this team. William Sr. had worked with his father (Thomas Leighton) under Jarves and William Libbey at New England Glass, and pioneered ruby glass. William Jr. had also worked with New England until 1863, when he helped build Hobbs, Brockunier & Company in West Virginia. George was the glass chemist at Hobbs, Brockunier & Company. Other managers and glass men (including the plant manager Bill Russell) were hired from Hobbs, Brockunier & Company. Dalzell, Gilmore & Leighton were making excellent pressed glass pieces such as lamps to compete directly with Libbey cut glass.

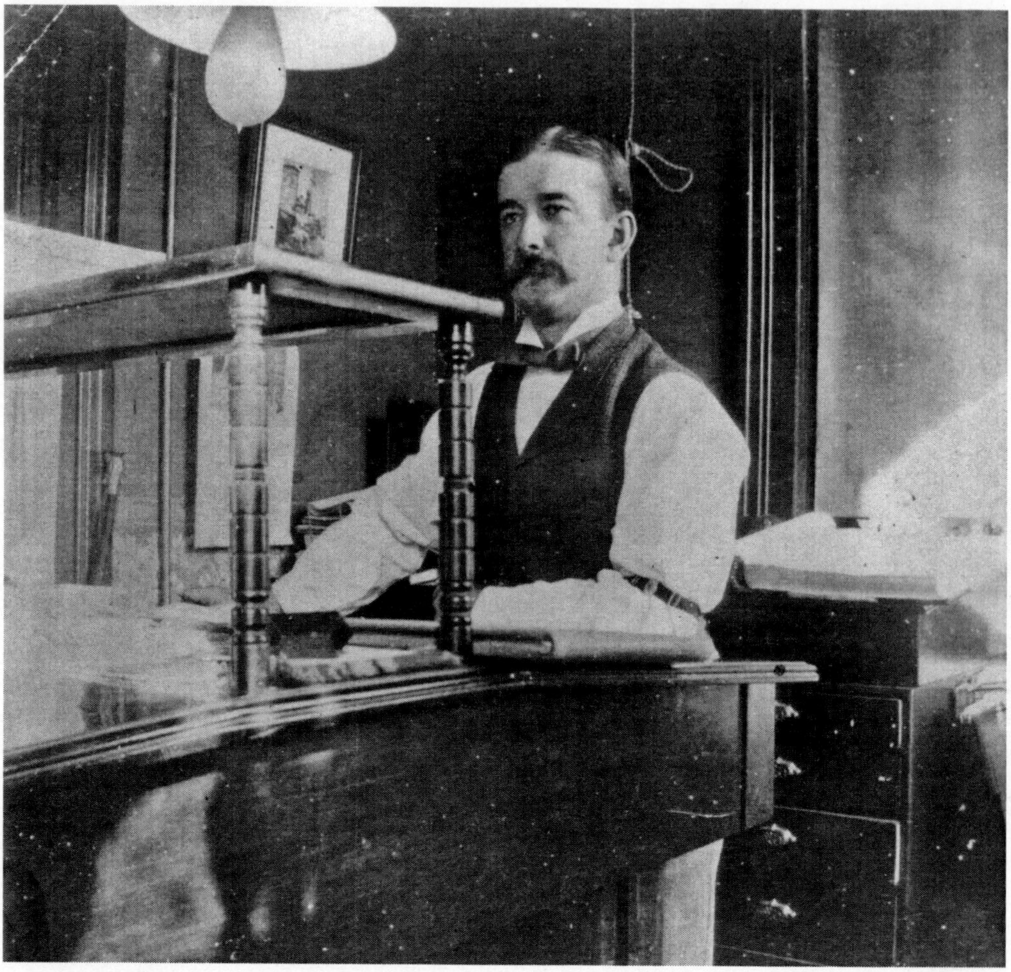

William Donovan, 1889 (courtesy of University of Toledo, Ward Canaday Center, Libbey Glass Records).

The new Findlay glasshouse of Dalzell, Gilmore, & Leighton Company broke ground in May 1888, before the Libbey Toledo plant was completed. The Dalzells moved their Wellsburg, West Virginia, plant to Findlay. The glasshouse was to have three 11-pot furnaces and cost $60,000, dwarfing the price of Libbey's plant in Toledo. By September 1888, the glasshouse was in operation, producing high quality tableware and lamps, although almost all in pressed glass. The Leightons contributed a variety of colors for the products. The company was considered one of the best, pressed glass manufacturers in the world. It was able to compete with some of Libbey's cut glass lines. In the end, however, Libbey's decision to locate in Toledo would prove to be smarter.

Chapter 7

The Turnaround

The story of Michael Owens is fraught with corporate mythology and legend. It appears that Libbey's secretaries worked on the story in preparation for a corporate history. This was to be a history like that of Carnegie Steel with epic stories. There is some evidence that Libbey kept the related historical documents in a safe, and it is rumored that his lawyers destroyed them at his death. Glass historian Warren Scoville saw the documents in the safe in 1921 and reported seeing court testimony by Owens given in a lawsuit against Owens Bottle.[1] Scoville said that Owens was made foreman of the blowing department after three months, and shop superintendent two years later. This would be consistent with Owens's experience and Libbey's management style. Moving Owens to foreman was a big step, but he was aggressive enough to push for the job. Owens would have been overseeing several "shops" or crews of glass makers. The full details of Owens's rise and early work are made less clear by the various versions that have been published over the years by Libbey employees.

By November 1888, the operation was in serious trouble. For three months, Owens had been complaining about the floor management as well as the inconsistent furnace operation. The furnace probably had a design flaw, but the discipline problem was even more pressing. Owens wrote Libbey notes about it, and Libbey was well aware of the issues. Owens had a temper and a big ego, but he was a knowledgeable glassmaker. He had worked every job in the glasshouse. Owens was not sure of the root of the furnace problem, but he knew how to take care of the discipline and training issues. Hopkins and Joseph Locke did not like Owens's complaining, but they respected his opinion. While outspoken, Owens was respectful when dealing with Libbey, always referring to him as "Mr. Libbey," a practice he is said to have continued his whole life. Libbey had to do something, and promoting Owens to foreman made good business sense. Libbey realized that Owens would clash with his plant superintendent, Locke. Locke was the antithesis of Owens. Locke had initially been described by *The Toledo Blade* as, "genial, popular, and obliging.... He is a talented man in other lines, being an artist and a musician."[2] Locke was every bit an artist, capable of producing etched and engraved pieces by his own hand.

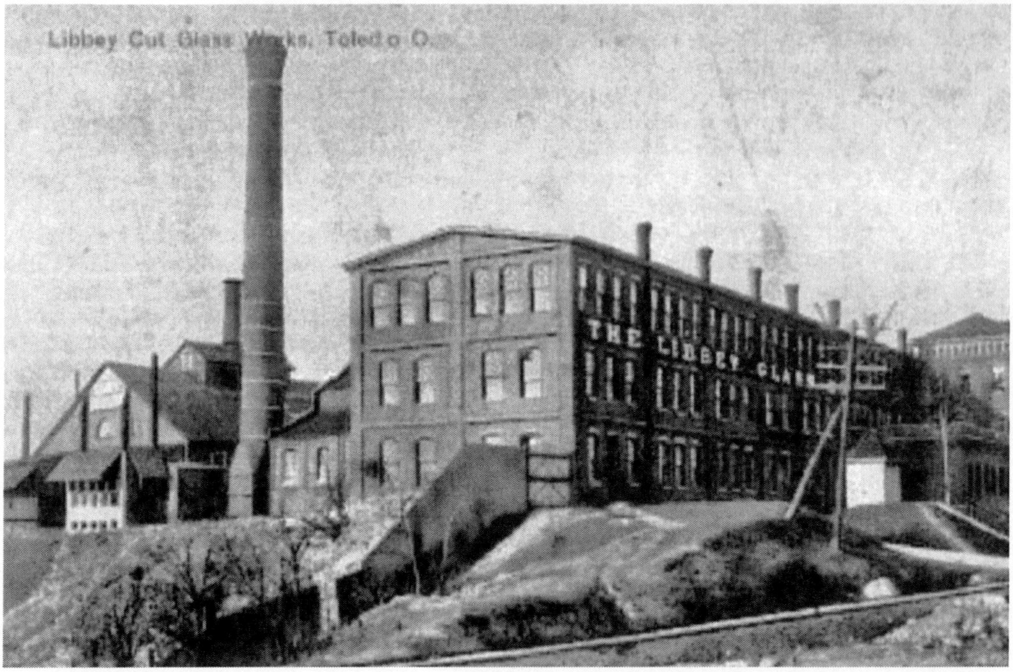

The Libbey Glass Plant, 1895 (courtesy of University of Toledo, Ward Canaday Center, Libbey Glass Records).

Personnel problems resulted in batches not being prepared on time. Owens's impatience and frustration grew, as did Libbey's. One story has Owens going to Libbey's office and giving two weeks' notice. Libbey executive John Biggers related this story years later.[3] Libbey was said to have offered Owens the foreman job after a long discussion. He had taken a liking to Owens, and there is no doubt that something had to be done. Owens's popularity with the men and his union background had to have played an important role in Libbey's decision. Libbey needed someone to be tough and discipline the workers. It is likely that Libbey's right-hand man, Solon O. Richardson, agreed that a change was needed. The story continues that Owens went on to fire everyone and then hire each person back after he committed to harder work. Owens confirmed the story years later.[4] The problem workers had to be let go, and Owens had the management sense to know it. With his old union ties, Owens knew he could replace them quickly with West Virginian glassmakers. The next thing Owens did was to apply tough rules and discipline for the factory boys to follow. He also implemented a clear training program for the boys based on his own experiences. By the end of the year, things were improving, with better morale and operating procedures. The furnace problems continued, but Owens now had the right men to take on the problem solving.

Locke and Owens had different ideas on how to resolve the furnace problems.

Joseph Locke had been involved in the original furnace design and even had a patent on the use of "producer gas" for glass furnaces.[5] Locke actually experimented with coal-made producer gas to run the East Cambridge furnaces. Owens and Locke clashed over the management of the furnaces. Owens showed little respect for Locke, and the rift became a problem. Libbey took a hands-off approach, allowing Owens to work things out with Locke. Libbey realized that he needed both men, avoiding the natural instinct to take sides with one. Owens was known to have many arguments with Locke, but Owens had the support of the crews. The furnace problem was slowly resolved, but Owens's floor management seemed to have immediate results.

Some good news in 1888 was that Libbey's gas bill came in at $4,246, a little over the estimate, but at the very low rate of 3 cents per 1,000 cubic feet versus the Pittsburgh rate of 7 cents. Owens started to work with the furnaces, and Libbey got the furnace manufacturers involved as well, which may have made the difference. The actual problem was never clearly identified, but the problem was with the lead flint glass product. This suggests the atmosphere controls were the problem. If the combustion gases were not vented correctly, carbon could combine with lead and precipitate out, clouding the glass. Temperature control could also be a problem. At the time, temperature control was a judgment call, and natural gas required experience. Locke had started the popular Maize glass back in New England and it required the skill of the new gangs to develop. Owens would spend hours experimenting to develop the needed expertise. Owens now had discipline among the workers but needed to nurture experience. He preferred to fill the crews with his own men. He was single at the time, living at a boarding house on Toledo's North Erie Street. Owens immersed himself in working at the glasshouse. By the end of 1889, the West Virginians outnumbered the Bostonians.

During most of 1889, Libbey saw a slow but steady improvement in overall operations. Owens loved to set challenging production goals and push the crews. He used his forceful personality and handpicked gang members to ignore union production restrictions, some of which he had developed as a union officer in West Virginia. Owens not only shortcut union rules, but he also cut administrative red tape. He took the time to instill a work ethic in the factory boys to assure smooth support operations. Deliveries and quality improved, but the price competition in the market increased as the glass industry overexpanded. By mid-year, the Maize glass was being delivered, and quality had returned to most of the cut glass. Solon O. Richardson was now convinced that Libbey Glass had to expand into new markets. Libbey was reluctant because of his personal commitment to high-end cut glass. Libbey's Chicago agent and friend, Jefferson Robinson, wanted to move into more utilitarian goods for the western market. Robinson started to spend more time in Toledo developing new product lines.

The tableware depression had started by late 1889. Libbey Glass started to consider other blown products such as bottles, oil lamp chimneys, chemical ware, tubing, and electric light bulbs. The company increased tumbler production, but pressed glass prices had been dropping. Tumblers fell from 55 cents a dozen in 1874 to 37 cents per dozen in 1888.[6] There were similar price drops for pressed goblets and wine glasses. Libbey lacked the equipment to be fully competitive in the pressed glass tableware market; however, Locke's blown Maize glass bucked the downward trend. The milk maize glass with stained green husks became a best seller in the mid-price market. It was relatively cheap to make, not requiring lead oxide or cutting. As a result, in 1889 the plant had only a $3,000 loss. Furthermore, Owens had demonstrated his ability as a factory floor manager.

Owens was able to expand his authority beyond that of a foreman. At heart, he was an operations man. The crews were now his men. Owens circumvented plant superintendent Joseph Locke, who was more interested in the art of glassmaking. Libbey demonstrated an amazing amount of trust in Owens. For months, Owens had to go to the Secor Hotel where Libbey was staying to get the mix instructions. Libbey had followed the old tradition of maintaining the batch (formula) book. Even plant superintendent Joseph Locke had limited access to the book. Owens convinced Libbey that he needed the book to fully study the glass furnace problems. Owens realized, as did Libbey, that the furnace problem was as much chemical as mechanical. In a real demonstration of trust, Libbey allowed Owens access to the batch book. To some degree, Libbey's social life probably played a role in his willingness to delegate. This move, however, increased Locke's distrust of Owens, and the relationship would only worsen with time. Libbey also started to delegate some of the paperwork by bringing his old friend, Jefferson Robinson, from Chicago to become a director and secretary of the company.

Libbey worked on the financial end and was able to obtain small corporate loans to meet payrolls. He was a personable gentleman who quickly assimilated into Toledo's high society, with two very important local assets — his lawyer, Clarence Brown, and his girlfriend, Florence Scott. Among his banking relationships was one with John B. Ketcham (1832–1897), who had opened Ketcham National Bank to take advantage of the glass industry expansion in Toledo. John and his brothers, Valentine Jr. and George, also had significant holdings in First National Bank, which their father, Valentine H. Ketcham, had founded in 1863. Both banks had a stake in the success of Libbey Glass. Libbey, however, was always very conservative with his loans. His loans through 1890 probably totaled less than $10,000, but they allowed him to pay bills and meet payroll. Libbey had an important relationship with the Ketcham family. John Ketcham, in particular, shared Libbey's and the Toledo Business Men's Association's vision for the growth of the glass industry.

Libbey Glass was restored to profitability in 1890 as the furnace operation was brought under control. The improvement continued all year, relieving Libbey of the daily worries of meeting payroll. In fact, Libbey Glass had profit of $75,000 by the end of 1890. Still, the tableware depression and overexpansion was taking a toll on the whole industry. But Libbey was able to maintain market share in the high-end cut glass market, which was spared thanks to the work of an Ohio Republican congressman, chairman of the Ways and Means Committee, William McKinley.

McKinley was a frequent visitor to Toledo, even though the Canton, Ohio, area was his district. Libbey first met McKinley at the newly incorporated Toledo Club in 1890. Carl Spitzer of the West End was a personal friend of McKinley and often had local dinners for him. McKinley had fought his whole career for protection of the American glass and steel industries. He pushed through his signature Tariff Bill of 1890 even though it cost him his seat in Congress, as the Democrats targeted him nationally for defeat. The protection saved the American cut-glass industry, which was under attack by aggressive foreign firms. Foreign workers in the cut glass industry were making up to four times lower wages. Unlike in the pressed glass industry, labor in cut art glass was up to 90 percent of the cost in 1890. European countries were offering incentives for exporting products to America to increase employment. The situation was so critical that many cut glass makers were actually considering moving to Europe. In addition, the bankruptcy of Bakewell Company, America's oldest and best known cut-glass manufacturer, had sent shock waves through the industry. The tariff bill maintained a 45 percent tariff on many forms of glass but increased the tariff to 60 percent on art and cut glass.

Without the support and protection of William McKinley, it is doubtful that Libbey Glass would have survived 1889. In 1898, J.D. Robinson took a beautiful cut bowl to Washington to present to President McKinley. Prior to being taken to Washington, the bowl was displayed for weeks in downtown Toledo. The bowl was used at the White House Thanksgiving dinner made headlines in newspapers across the country. It would endear McKinley to Northwest Ohio, and years later, shortly after his 1901 assassination, Toledo would be the first to raise a monument to him, which still stands today in front of the courthouse. Libbey and the Toledo capitalists were strong supporters of McKinley and his policies. McKinley's vice president, Theodore Roosevelt, made a special visit to Toledo's wealthy West End in 1900 to thank its residents for their support. Libbey's friend and neighbor Carl Spitzer arranged the visit. Spitzer would travel with McKinley weeks later to christen the battleship *Ohio*.

The McKinley tariff and Michael Owens gave Libbey the breathing room he and the company needed. Libbey felt he was now in a position to marry Florence. The wedding, which *The Toledo Blade* called the social event of the season, occurred on June 25, 1890.[7] It was clearly the merging of the old Toledo aristocracy with the new aristocracy. The wedding took place at Maurice Scott's new home at 2449 Monroe Street (the future site of the Toledo Art Museum). The house and grounds were lit up the night before in a beautiful display. The Scott House had just been built in

1889 in a park-like setting of native trees. The house was then the center of the fashionable West End subdivision developed by Florence's uncle and son of Jesup Scott, Frank Scott, who would become known as Toledo's West End architect. Frank Scott was Toledo's first architect and developed the West End, which would be the future home of many Libbey Glass executives. This beautiful neighborhood embodied the ideas of Frank Scott's 1870 book, *The Art of Beautifying Suburban Home Gardens*, the first book about suburban planning. Brother William Scott resided next to Maurice where the West wing of the Toledo Museum is today.

The guest list included many residents of the West End such as Rudolph A. Bartley, Toledo's wealthiest grocer; William Brigham, banker; John Ketcham, banker; and Christian Gerber, Toledo's wealthiest merchant. Dr. W. Williams of the First Congregational Church performed the ceremony. Jefferson Robinson was Libbey's best man and Miss Genie Moulton was Florence's bridesmaid. There was an extensive orchestra and refreshments for Toledo's wealthiest residents. Afterwards, Libbey and Florence left for a short East Coast (Boston and New York) honeymoon. They would return to take up residence at 112 Twelfth Street, at Madison Avenue.

Tariffs and Mike Owens had the company on track, but there were new challenges. In late 1888 and early 1889, the first signs of a limit to the natural gas supply appeared. Houses noticed a loss of pressure in November of 1888, and then the gas failed in December 1888. On December 18, 1888, Libbey Glass lost gas supply for a day and had to use wood to fire the furnaces. The company had to prepare for gas failures and this would continue for years. Libbey's idea to use multi-fuel furnaces now started to look like genius. The good news was that gas shortages finally ended the rush of glass companies into the area. The gas company responded by more drilling to hold off a crisis. The Findlay, Ohio-based gas industry ended free gas by 1891. Glass companies in Northwest Ohio started to fail, while others packed up and followed the cheap gas to Indiana. Many filed lawsuits against the city. Larger Findlay companies had to pay a new rate, which took the gas bill of Dalzell, Gilmore & Leighton to over $5,000 a month. This and the drilling helped stabilize the gas supply. With a firm price contract, Edward Libbey became a preferred customer, but he started to back up his operations with oil and coal. Oil for a time was much cheaper, and Libbey's multi-fuel furnaces were able to take advantage of the price decline.

There were opportunities as well. One of the growth segments in the glass market was electric light bulbs. Edison General Electric and Corning Glass controlled the electric bulb market. Libbey Glass had some electrical business. Richardson and New England Glass were a major producer of electrical insulators; in fact, Joseph Locke even had a patent for electrical insulators. Corning Glass Works of Corning, New York, controlled 95 percent of the production of bulbs for Edison General Electric. Incandescent light bulbs at the time were hand blown into a mold. The gang consisted of a gatherer, blower, assistants, and several boys, including a mold boy who operated

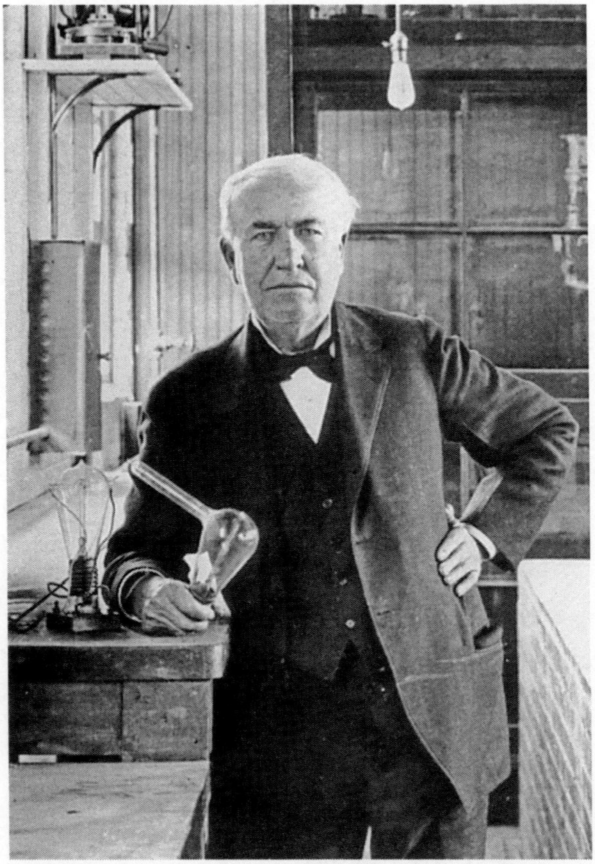

Libbey developed an early relationship with Thomas Edison to make labware and light bulbs. Edison shown here in his lab ca. 1910 (courtesy of University of Toledo, Ward Canaday Center, Libbey Glass Records).

the mold by hand. A gang of six at Corning could produce two bulbs per minute or about 1,100 bulbs per gang per shift. This type of repetitive hand manufacture was the hardest work in the industry. A struggling Edison Electric applied enormous pressure, forever increasing production rates. Thomas Edison and his company were highly leveraged and wanted suppliers such as Corning to bring down prices. Edison was slowly losing his company to bankers such as J.P. Morgan. Corning Glass, at the same time, was facing the upward wage pressure from the American Flint Glass Workers Union. Corning was an open shop (non-union), so the battle was also over recognition of the American Flints. An open shop did not require workers to join the union. The 360 workers at Corning went on strike in December of 1890 for wages and union recognition. Edison General Electric started discussions with its old friend, Solon O. Richardson, who had worked with Edison Tubing to make crude bulbs in the early days.

Libbey Glass immediately became involved in the crisis. Edison General Electric needed a supplier quickly because of the demand for electric light bulbs. The American Flints and the workers had to be careful not to lose business due to the strike. As the strike moved into mid–January, both Edison General Electric and the American Flints were courting Libbey. Edison General Electric needed a supplier fast. The demand was for over a half a million bulbs a month and growing every day. Each day of lost production was a crippling blow to Edison General Electric's business. Libbey Glass was the natural choice to supply Edison General Electric. Solon O. Richardson had ties going back to Edison's earliest days. Libbey Glass also had maintained a small amount of bulb production over the years. Libbey Glass had the experience and glassblowers to move into production quickly. The union at Corning was becoming anxious that the business could be lost forever if Edison contracted with another supplier. The union approached Edward Libbey with a proposal. They would supply all the additional blowers needed and suspend many union production restrictions. For his part, Libbey would pay normal union wages (which were much higher than those at Corning) and promise to give back the business to Corning when the strike ended.

Libbey was under pressure to make a decision by the third week of January, 1891. While he was the major stockholder, Libbey did have an advisory board of Solon O. Richardson, Jefferson Robinson, Joseph Locke, and William Donovan. Richardson, of course, was a major supporter of the deal, as was most of the board with the exception of Locke. Libbey hesitated because he had several concerns. First, Libbey had just got his art and cut glass business on its feet. Bulb production would take up most of the capacity of the Toledo plant. Libbey had just struggled through a disruption in the employee morale and internal divisions between the New Englanders and West Virginians. Gas pressure problems were limiting production at Toledo. Libbey finally agreed to take the Corning contract and the union deal in the third week of January. But he planned to produce the bulbs at a deserted glass plant in Findlay, so he could maintain art glass production and avoid internal scheduling problems and gas pressure problems. Richardson was dispatched to Findlay to rent the old Globe and Chimney Works plant.

The lease was signed on January 23 for 18 months at a cost of $1,800 a year. The plant had only been idled two months earlier. The Globe factory had a large, 16-pot furnace, which could meet the light bulb demand. It had closed because of a lack of glassblowers. The Corning union quickly moved to supply 150 workers and had 50 on site by January 27. Libbey gave Richardson full authority for the project, but he moved Owens to Findlay to run the factory. It was a strange mix of personalities, and even though both were relatively young for such an undertaking, Libbey made the right call. Both men knew the customer and the product better than anyone in the organization. Owens had demonstrated his ability to manage the factory floor and had been an officer in the American Flints. Owens's turnaround of the Toledo plant had been nothing short of a miracle. Richardson was an organization man and a follower of the scientific management of Fredrick Taylor. Owens was a tough, natural, goal-oriented manager who would often take administrative and organizational shortcuts. The pressure to start up the factory without problems would require both men.

The factory went into operation on February 2 with Edison Electric already desperately in need of bulbs. That required a 24-hour operation with a suspension of any union work rules, allowing Owens to take all kinds of liberties. Glassblowers at Corning made $12 a week while at Findlay they were making up to $28 a week.[8] Boys were paid double what Corning paid. Owens started with the Corning workers and brought in West Virginians, as boys were in short supply in Findlay and Northwest Ohio. Owens and Richardson both moved to Findlay and worked long hours to get the operation moving. By the first month of operation, the plant was profitable. Production rates were over a quarter million bulbs per week. General Electric found the quality of the bulbs exceptional. Richardson got credit in the press, which widened the dislike between him and Owens.

The operation survived a major tragedy in July 1891. The Findlay bulb plant had been running 24 hours a day, but even with amazing productivity, it was fighting a backlog of orders from Edison General Electric. The men decided to forego the normal six-week summer shutdown for a shorter two-week hiatus. A group of 19 glassworkers returning to Corning for the short vacation were killed in a train wreck at Ravenna, Ohio. It was a real blow to the workers and community. Today, a memorial built by the American Flints can be seen at St. Mary's Cemetery in Corning, New York.

The high productivity of the plant was directly related to several innovations of Mike Owens. The initial gang consisted of a gatherer to collect a ball of glass at the end of the blowpipe. The blower then took over. A mold boy guided the blown bulb into a mold. Another boy cracked off the blowpipe when completed. Then the anneal boy took the bulb to the anneal, while yet another boy cleaned the blowpipe. Owens himself had been a mold boy and well understood the process. The mold boy often was the key to the speed of the operation. Owens used a paste mold system where the wooden or cast iron mold was coated with carbon and linseed oil for quick release. The paste was a mixture of wax, oil and carbon, which Owens had developed through experimentation. Paste-molds had been experimented with in West Virginia, but Owens perfected the process at Findlay and would eventually apply for a paste-mold process patent.

Late in the summer of 1891, the mold boys threatened a strike. The boys were not represented by the American Flint Glass Workers Union; in fact, the union took no note of the boys' problems. Mold boys were often from eastern orphanages and were crammed into overcrowded boarding houses. Their complaints were not so much for more money as for better living and working conditions. Richardson and Owens could ill afford a slowdown in work. Owens had long been considering an automatic release of the mold as a means to speed up the process. The threat of a strike only encouraged Owens to look more seriously at such a device. Owens, however, was not a mechanic. He had drawn the device and envisioned its action, but he was probably incapable of fabricating it. Every glasshouse had a blacksmith shop which did fabrication and repair. Owens teamed up with plant blacksmith James Wade to build a crude, foot pedal-operated device. Solon O. Richardson maintained that it was one of his ideas that made the Owens machine practical.[9] Owens was able to get many of these devices working, which greatly increased production and could eliminate

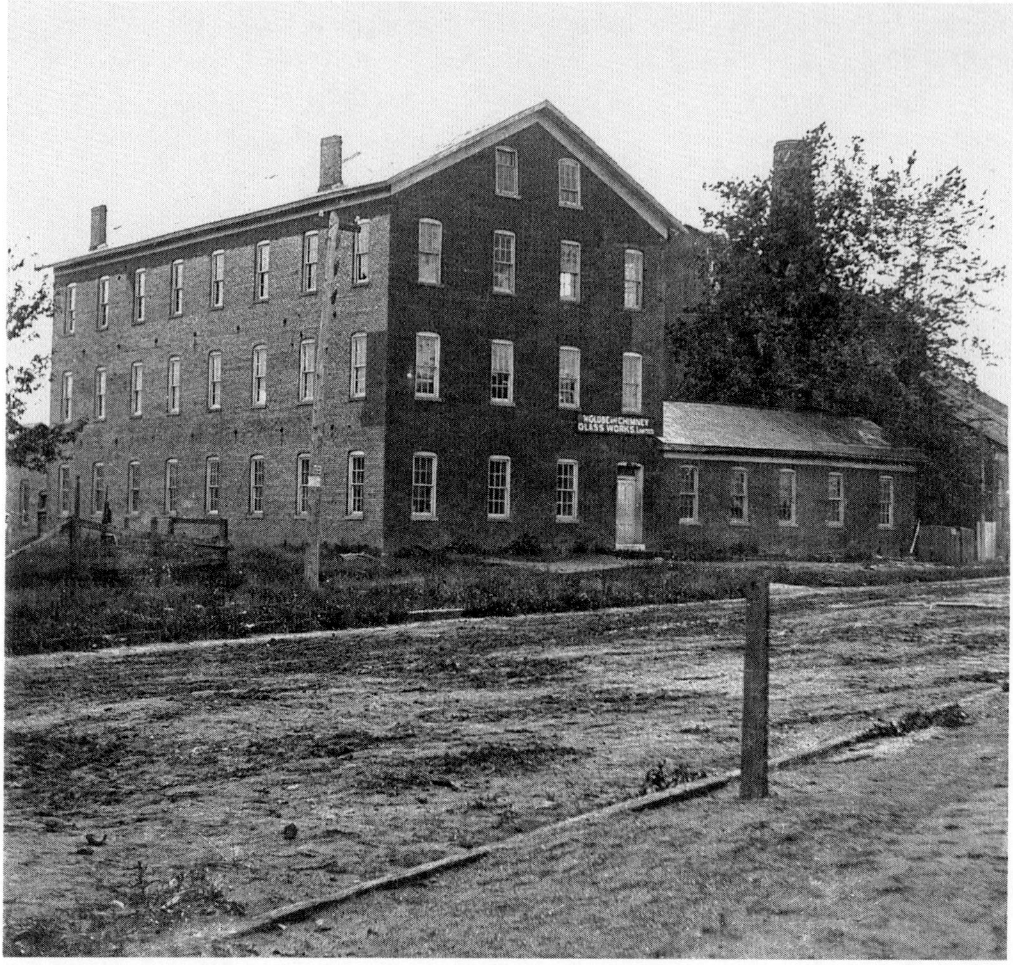

The plant that Libbey rented in Findlay, Ohio, in 1891 for light bulb production (courtesy of Hancock County Historical Society).

one boy from the crew. Fortunately, the boys never struck, but Owens did increase productivity with his device. This would be the beginning of automation of the glass industry. American Flint officers made several visits to the new operation and were impressed with the higher wages and better morale, even with Owens often ignoring work rules.

Owens seemed to fully enjoy being factory manager. A devout Catholic, he made friendships at Findlay St. Michael's Parish which he maintained throughout his life. Certainly, Findlay would miss Owen's presence in the church and community. The rift between Richardson and Owens deepened as Richardson got all the publicity for the amazing success at Findlay. But Richardson was also well-liked in Findlay. There were rumors that Libbey might move his whole company to Findlay, but these were nothing more than rumors. As the plant moved into the winter of 1891–1892, it had gas pressure problems, as gas shortages hit the Findlay area. But Richardson was convinced that Findlay would end operation in the summer, assuming Corning Glass could take over production again. Libbey had already started a major expansion in Toledo with a ten-pot furnace for bulb production. But gas pressure had dropped there as well. Libbey started the conversion to oil in January 1891 and had to struggle again with furnace problems. He negotiated an outstanding oil deal with Standard Oil of Cleveland and converted to oil in the spring of 1892. Problems continued into the fall of 1892 as Libbey had to revamp his furnaces again. A compressed air oil spray system finally corrected the furnace problems. Now, with the oil working at a low price, Libbey was in the best fuel position of any Northwest Ohio company.

Libbey Glass was considered the area's low-cost glass manufacturer thanks to the guidance of Edward Libbey. The Findlay bulb operation produced a huge cash flow, allowing Libbey to pay off all his debt. Early in the project, Libbey saw the wisdom of expanding into the bulb business and started work on a new furnace in Toledo. He had been hesitant to mix his art production with more utilitarian glass production, but the Findlay natural gas shortages of the winter of 1891–1892 convinced him that he needed to invest in an oil convertible furnace. Richardson wanted to bring bulb production to Toledo. Mike Owens, on the other hand, enjoyed the separate plant in Findlay. Toledo again had the advantages of cheaper fuel and women needed for packaging and shipping. Libbey only wanted to make the most profitable decision.

Chapter 8

The Columbian World Fair

Bulb production at Findlay ended in July of 1892, but Libbey Glass had been changed forever. Libbey was producing light bulbs in Toledo and was an established second supplier for Edison General Electric. He now had the perfect product mix for economic downturns. The company was highly profitable and flush with cash. By the beginning of 1893, Libbey reported a $109,500 profit. The company gave a 10 percent cash dividend and changed its name to Libbey Glass Company. Solon O. Richardson was rewarded with a $10,000 stock bonus and a substantial salary increase of $3,000. This block of stock made him one of the larger stockholders. Richardson was also named general manager of operations. Joseph Locke had left the company, probably realizing that Mike Owens was going to be made Toledo plant superintendent. Libbey felt he had no choice but to promote Owens, realizing he was a bit of a maverick. Libbey Glass, however, faced more challenges in 1892. The country was entering a deep depression. Competitors formed a combine of 18 companies by mid–1892, known as United States Glass Company ("Flint Glass Trust") and specializing in tableware and art glass. Joseph Locke had gone to the new combine as chief designer, which further concerned Libbey. Pittsburgh financiers backed the new trust. But Libbey Glass was America's most profitable company thanks to the executive skills of Edward Libbey. Libbey was starting to enjoy the income.

United States Glass Company was America's first huge corporation and would be a model for the formation of United States Steel in 1901. United States Glass Company came into being because of industry over-expansion and the tableware depression of 1890. Small companies lacked the brand name power of Libbey Glass. Some were overwhelmed by spiraling costs as competition drove prices down. As early as 1878, flint glass manufacturers had started to pool resources to purchase raw material and to stabilize prices. Edward Libbey had always refused to join these pools. Libbey always believed that free competition was better for the industry and his company. One of the earliest manufacturer associations was the Western Flint & Lime Glass Protective Association, which Libbey had refused to join in 1878. This organization had some of the major names in pressed glass such as Ripley, Bryce, Adams, Brockunier

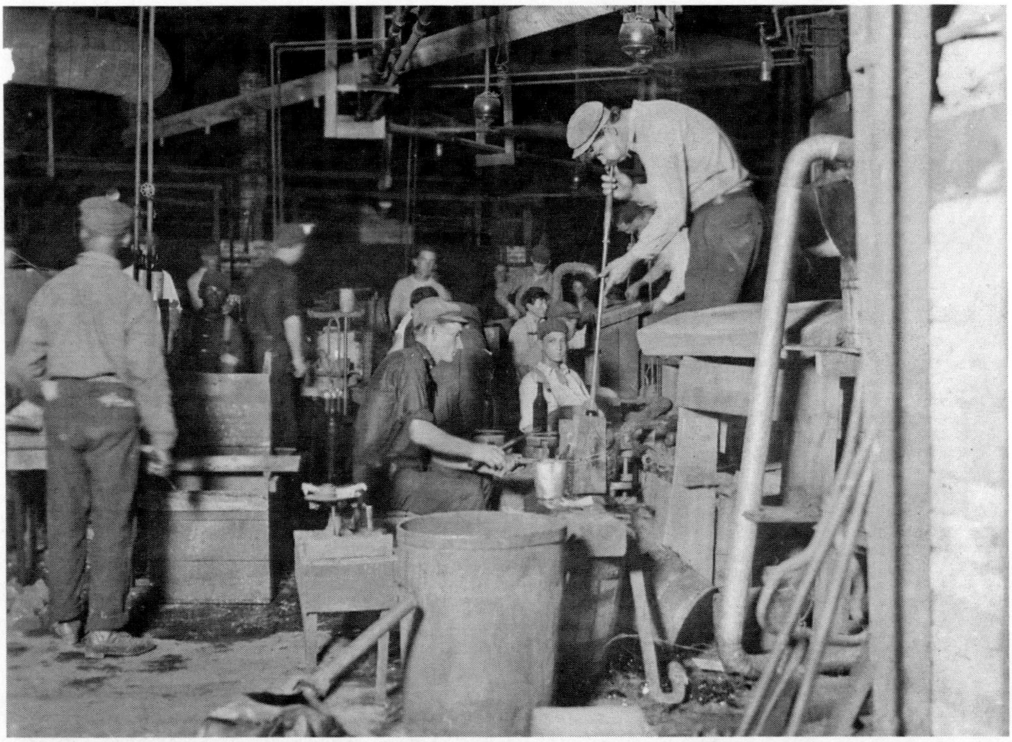

Libbey Glass blowing operations, 1895 (courtesy of University of Toledo, Ward Canaday Center, Libbey Glass Records).

and Gillinder. Pittsburgh glass mogul Daniel Ripley (1850–1912) had been its president. The Western Flint group was rendered superfluous by the glass boom of the 1880s. By 1890, the tableware industry was on the verge of a major shakeout and potential collapse. Libbey's high-end product and opportunistic bulb production had allowed him to escape the damage of the tableware depression and industry overexpansion.

In 1890, as the tableware depression reached its height, Daniel Ripley started the process of building his consortium. These small glass company owners were under extreme cost pressures, and many were facing bankruptcy. The meetings continued into July of 1891, when United States Glass Company was formed in Pittsburgh, bringing together thirteen factories in Pennsylvania and Ohio. While the factories would operate under their own management, the corporate board would control marketing and overall product lines. Unfortunately, it was a merger of necessity not a market strategy. Initially, this huge combine threatened Libbey, but United States Glass had a mix of union and labor problems. Libbey continued to target the high end of the market. He had already converted to oil when the natural gas shortage reached a crisis level for many of United States Glass's Findlay area plants, driving up costs and forcing shutdowns. Libbey's approach was to move out of the pressed glass market, which seemed locked in a depression of its own. He observed that glass tariffs were under attack as the Republicans and William McKinley were swept out of office in 1892. Without tariff protection, the pressed glass market would be under heavy pressure from cheap European products. Libbey saw the best market for the future as the cut glass and art glass market. Bulb manufacture could help cash flow because of its high volume and high profit margin.

For a short period at the Toledo plant, Owens and Richardson contended for leadership. Owens felt his role in the success of the Findlay bulb plant had been minimized by both Richardson and Libbey. For his part, Richardson believed he had made Owens's paste-mold system operational. Libbey remained neutral. As a concession, Owens was allowed to sell the rights to use the machine to United States Glass Company to produce tumblers, receiving $5,000. Libbey probably realized that he would have to keep Owens and Richardson apart as best he could. Having lost Joseph Locke, Libbey did not want to lose either Richardson or Owens.

Libbey would always favor art glass production, and he had a project in mind after the Findlay success. A massive world fair was being planned to celebrate the 400th anniversary of the founding of America. The fair would eventually be a triumph of electric lights, with George Westinghouse winning the contract to light it up. The fair would need three times the power used by the whole city of Chicago. Almost 93,000 incandescent lights and 5,000 powerful arc lights would light the "White City." Most of these bulbs were made at Westinghouse's glass plant because of their unique shape. The dominant symbol of the fair was to be fellow Pittsburgher George Ferris's invention — the "Ferris Wheel." The fair would have over 28 million visitors. General admission to the fair was 50 cents for an adult and 25 cents for children. The fair would be remembered for the introduction of Cream of Wheat, Juicy Fruit Gum, Pabst Beer, Aunt Jemima Syrup, Shredded Wheat, the Heinz pickle, and the hamburger. Libbey wanted to have a full glass factory built on the midway. The Chicago fair was meant to dwarf the Great Exhibition of 1851, the Philadelphia 1876 Exposition, and the Paris Fair of 1889. Libbey wanted not only to be the premier glass company at the fair, but the premier company in general, remembering his experience as a youth at the 1876 World Fair at Philadelphia. It would be a launching pad to establish Libbey Glass as the world's greatest cut glass manufacturer. But no one knew how successful the fair would be when Libbey proposed his plan to his advisory board in February of 1892.

The board members, Jefferson Robinson and Solon O. Richardson, were less than enthusiastic about spending cash on the Chicago Fair. The year 1892 brought concerns for the future of the glass industry. A Democratic victory was being predicted, and Grover Cleveland was campaigning on reduced tariffs. Furthermore, the economy was showing signs of weakness. There were disarray in the money markets and problems on Wall Street, which concerned many on Libbey's board. The fair plan would require major borrowing. He planned a complete factory with a ten-pot furnace and over 100 employees. The factory would have a complete finishing department with cutting, engraving, and etching and would have room for 2,000 visitors. Libbey's more than $200,000 estimate had to be sobering to the conservative members of his board. Eventually, Libbey got $75,000. He formed a separate corporation of the Libbey Glass Company of Illinois. Libbey personally borrowed another $100,000 from banker John Ketcham. With the backing of Libbey Glass and Ketcham, Libbey borrowed another $100,000 from Toledo banks, who trusted him because of his demonstrated success.

He would need the entire $250,000 to implement his vision. Libbey proved skillful in his negotiation with the fair committee. His contract called for the exclusive right to sell glassware on the fair grounds, which cost him $5,000. This assured him a central location on the midway. Libbey hired Toledo architect D.L. Stine to design the beautiful Italianate pavilion on the Midway Plaisance. A large dome was designed to hide the smokestack required by the glass furnaces. The pavilion had living quarters on the second floor for Libbey and Florence. The building looked like stone, but like most of the fair's prestigious buildings, had a stucco facade.

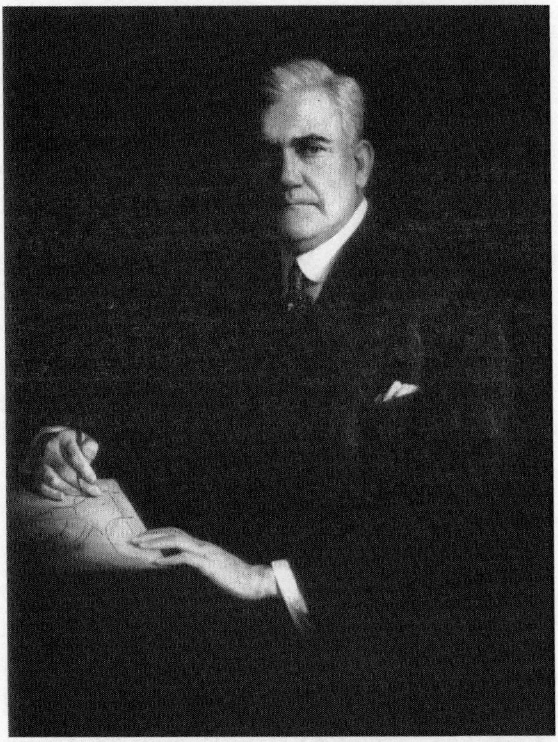

Michael Owens (courtesy of University of Toledo, Ward Canaday Center, Libbey Glass Records).

Libbey would also have a glass exhibit in the Manufacturing Arts Building with other glassmakers. His selection for the factory manager was Mike Owens. The assignment got Owens out of Richardson's hair and would allow him to continue experimentation with new processes. Libbeys best gaffer, Joseph Rosenberger, and his gang would amaze crowds with their exhibitions of glass blowing.

The fair proved to Libbey the importance of mass marketing. His exhibit would stand with that of H.J. Heinz for advancing the science of marketing. Heinz introduced his famous pickle pin at the Columbian Exposition. Libbey also introduced one of his personal projects to the world at the fair. Like any good marketer, Libbey loved new products. For months, Libbey had secretly been developing "spun glass." Not even Libbey Glass directors were aware of this operation. Libbey often chose to do product development outside his company. One of his business strategies was to develop products and processes and then sell them to his own company and others. This strategy seems to have roots in Deming Jarves's approach.

Libbey had brought Herman Hammesjahr from Germany to Toledo to develop spun glass, which was much like fiberglass today and could be made into cloth-like products. Hammesjahr actually combined glass fibers with silk. Libbey and Hammesjahr hired 45 girls to help manufacture spun glass products for the fair, such as neckties, napkins, and even dresses. They offered demonstrations of spinning from molten glass.[1] Each visitor received a spun glass pin as a souvenir. One marketing angle was to create a spun-glass dress for Georgia Cayvan, a famous actress of the time. Cayvan tried to wear the dress several times, but it was too brittle for regular use. The dress did attract fairgoer Princess Eulalie of Spain, who had a large entourage of press following her and her friend the Duke de Veragua, a relative of Christopher Columbus. The princess had a duplicate dress made, and Libbey became the official glasscutter for Spain. While spun glass did not take off, Libbey got extensive publicity in Europe and America.

Much has been made of the slow start of the exhibition. One story says that Mike Owens hired 250 "cappers" to flood the exhibition to create interest. Another is that a Libbey employee suggested charging an entrance fee. Originally, the Libbey Pavilion was free; an admission charge would make the exhibit appear more desirable. Therefore, a charge of 10 cents was imposed. The charge was increased to 25 cents to help cover operating costs when early crowds were thin. As the crowds grew, the price increased to 50 cents to limit numbers. The price, however, could be applied to the purchase of a glass piece. H.J. Heinz Company has similar lore about the first month of the fair. There were many problems including the weather during the opening month of May, with rain and muddy streets. The low attendance created a near panic in several large corporate exhibitors. Many did indeed hire fairgoers to hang around their exhibitions. It was so bad that the fair committee had to let go 3,000 of 6,000 employees.[2] Attendance began to pick up dramatically in June, July, and August, blooding all the exhibitions.

Facts suggest a different a story than the corporate love. The Libbey pavilion daybook shows from 35,000 to 40,000 visitors a week during May.[3] Neighboring attractions complained about Libbey's free show. Libbey imposed admission on May 15 of 10 cents, increasing it at the end of the fair to 25 cents. Considering over two million visited the pavilion, this amounts to a large sum of money. The admissions could have meant as much as $300,000 to Libbey, which would have covered his costs. Logbooks show that sales in the crystal room ranged from $900 to $4,000 a week.[4]

Libbey's glasshouse was a world-class exhibit. He brought his best blowers and glasscutters to the fair and he cut some of his finest punch bowl sets to date there. Willie Anderson designed a special pattern known as Kimberly. Libbey had a crystal art room made of mirrors and electric lights, which caused the cut glass to take on a diamond-like quality. He commissioned a popular author, Kate Field, to write a history of glass. He sold souvenir books of the Libbey exhibit and special books for children. Numerous souvenirs were available. The U. S. Postal Service had introduced the "postal," a picture pre-stamped postcard. Libbey printed cards with his exhibit on them that mailed for one cent, half the price of other mail. He offered a full array of glass pieces such as tableware, cut glass, and paperweights. At one point, Libbey ran low on products and went around the country to purchase Libbey products to bring to the fair. The fair consumed him. His interactive and hands-on displays were considered models for future fairs.

Both Libbey and Owens spent considerable time at the fair. Libbey was able to work with an old customer, Louis Tiffany, and found a new one in H.J. Heinz. For Owens, the fair was as inspirational as the 1876 Exposition had been for Libbey. One story claims that Owens visited the Chicago slaughterhouses and got the idea for an automated glassblowing machine. (The story is similar to one told about Henry Ford visiting Cincinnati slaughterhouses and getting the idea for the assembly line. Owens's eventual bottle-making machine was closer to the slaughter house mechanism than Ford's assembly line. Such stories promoted corporate culture and mythology in the early twentieth century.)

Owens might have visited the slaughterhouses, but he was well on his way towards glass blowing automation at the time. He took time to experiment with a number of projects such as his automated mold. Libbey clearly encouraged this type of development by Owens. Some work was done on the automated paste mold machine for light bulbs. Thus Libbey's investment in the fair paid development dividends. Owens may have started his work on automated bottle blowing. All of this would have been done without the knowledge of Libbey Glass's board of directors. Libbey preferred to work on major development projects outside his companies to assure control and reap the most personal return on the dollar.

The fair turned into an amazing financial success. Libbey paid off all of the loans and fees and made a $100,000 profit. The company reported a profit of $147,026 for 1893. Libbey was in an ideal cash and product position to weather the depression the country was entering. Just as important was the publicity and marketing advantage that Libbey had envisioned and gained. He had developed a friendship with H.J. Heinz at the fair. H.J. Heinz and Libbey were America's greatest marketers of the period, and Heinz had a love of glassmaking from his boyhood. Heinz Company was one of the larger users of glass bottles and containers. Libbey was impressed with H.J. Heinz's use of branding to make his products the most asked for. Libbey also developed a closer relationship with Louis Tiffany, who had a neighboring exhibit. Libbey's engraved punch bowl set of scenes of the fair won a gold medal and international acclaim. He launched a major advertising campaign to publicize this acclaim and his success at the world fair. Profits continued through the recession years of 1893 to 1898 with dividends being paid from 40 percent to 100 percent. Libbey's cut glass was world renowned.

Chapter 9

Industrialist and Glass Aristocrat

The Chicago fair had been an enormous success and created quite a surge in Libbey Glass sales. On this foundation, Libbey launched a marketing campaign to brand Libbey glass as the best. After the fair Libbey took a breather and went on vacation with his wife in California. He was the owner of one of America's most successful glass companies. And even with the depression, he could count on the booming light bulb market and steady sales in high-end cut glass. With Democrat Glover Cleveland in the White House, the tariffs on cut glass were reduced from 60 percent to 40 percent, allowing a significant increase in cut glass imports. Even so, Libbey had the right organization to remain successful. With that, Libbey and his wife wanted to enjoy life more. He could start evolving into a true chief executive, allowing the organization to do the heavy lifting. He could continue his efforts in sales and marketing, and with Mike Owens, he could look to making money from the development of new processes and products. Libbey's closest business friend, Clarence Brown, was fast becoming his Toledo corporate lawyer. Brown had prior business experience with the railroads. With Brown's advice, Libbey felt comfortable discussing his problems with both Richardson and Owens. Libbey's older friend, Jefferson Robinson, had also grown to be a trusted friend of Richardson, so Brown offered a neutral sounding board. Owens returned to Toledo, not to manage but to pursue a breakthrough. Life for Solon O. Richardson as vice-president and general manager of Libbey Glass was about to get more complicated.

Mike Owens returned to Toledo with hopes of fully developing his paste mold process for light bulbs. Richardson felt the development work should be owned by Libbey Glass, but Owens and Libbey treated it as a separate project, even though Owens was using Libbey Glass resources. Owens was more a visionary than a mechanic. His crude drawings were three-dimensional, and like Thomas Edison, he needed draftsmen to make machine drawings. He had developed the process to the point where he required extensive engineering help. Owens, Ed Jewett, master mechanic of Libbey Glass; and Alex Kinnear, pattern maker at Libbey Glass developed a number of rough machines for testing. Successful tests led to assembly blueprints for production of the equipment. Richardson, at the same time, was pursuing a different approach to a bulb-making machine. Relations between Richardson and Owens worsened. Owens, in particular, could not tolerate supervision. Libbey would give Owens verbal approval, and Owens would ignore organizational structure to bulldog his development work.

In 1894, Owens had the drafting department of Libbey Glass produce blueprints for Toledo Machine & Tool Company. Owens then tested the machine at Libbey Glass, again over the complaints of Richardson. Libbey was getting an order from Diamond Incandescent Lamp Company. The semi-automatic machine proved successful, and Libbey helped Owens obtain a patent through his patent lawyer in Boston. The internal feud between Richardson and Owens turned bitter, and years later it would lead to a lawsuit. Richardson suggested that most of the work be done by Ed Jewett. Libbey realized that Owens was a creative genius but that he needed a lot of mechanical and engineering support. Owens was a creative inventor. Libbey was a lot like Owens in that he was results-oriented.

Libbey tried to stay out of the internal turf wars, not viewing the corporation from a modern standpoint. Because Libbey controlled almost 70 percent of Libbey Glass shares, he viewed it as a personal asset. This was not unusual for these earlier capitalists, but the minority shareholders did not see it that way. They believed that they should get at least an equivalent share of any developmental work done within the company. Libbey and Owens offered the free use of the inventions but also sold them to competitors. Libbey saw the advantage in selling rights and leasing his machines and processes.

The 1894 semi-automatic bulb machine was a major advance in productivity because of its speed. The machine rotated five molds on mechanical arms. The actual blowing was done by compressed air through blowpipes. The machine had to be fed by three men who gathered the molten glass on the blowpipes. Another man placed the blowpipes onto the rotating machine. No skilled workers such as blowers or gatherers were required. The machine also required six boys to remove pipes and bulbs, crack bulbs off pipes, and carry them to the anneal. The machine could make 2,000 bulbs in five hours, which would take five glass blower gangs working for five hours. Overall costs were reduced an amazing 90 percent. It was the industry's first step towards real automation. Libbey started to realize the potential of a line of glassmaking machines, and got excited about automation. He still had a passion for art glass, but bulbs and tumblers were not art. Light bulbs were, however, a growth market as America converted to electric lighting in the 1890s.

The years 1894 and 1895 were amazingly successful for Libbey Glass and Libbey personally. Most glass companies experienced a great national depression following the tableware depression of the early 1890s, and natural gas shortages destroyed many companies. Nearby Findlay cut off natural gas to factories as the city tried to save gas for home consumers when supplies waned. Northwest glass companies closed one by one. Findlay Window Glass tried a new producer gas system but failed in 1895. The use of producer gas was plausible, but the cost in the middle of a depression was too much for many of the companies. The United States Glass Company conglomerate continued to struggle with economic and gas issues. In late 1893, the American Flint Glass Workers struck, seeking union representation in the non-union plants. The strike helped close a number of plants permanently, extending on and off to 1897 and effectively killing the United States Glass Company.

Layoffs in the fall of 1893 were approaching a million nationwide, and 141 banks had failed. The smokestacks of the Mahoning Valley and Pittsburgh were cold. Two old steel companies — Pennsylvania Steel and Oliver Iron and Steel — went bankrupt. Railroads failed almost monthly, producing massive unemployment after over-extending. Unemployment was probably greater than 25 percent. Wage cuts were causing a rash of strikes as the country declined further into the depression. Unemployment produced unrest and upheaval. The year 1894 would be the darkest. National union leader Samuel Gompers estimated that three million were out of work. Historian Henry Adams said: "Winter is here and my perpetual miracle is that people somehow seem to go on living without work or money or food, or clothes, or fire. One or two million people are out of work; thousands of rich are cleaned out to the last shoe leather; not one human being is known to be making a living." Times were indeed tough. Even capitalists like Mark Hanna were feeling the pain. In Toledo, A.L. Spitzer's steel "skyscraper" had to be put on hold. Spitzer rushed to Pittsburgh to speak personally with Andrew Carnegie to stop steel shipments. Libbey's light bulb business and the popularity of its high-end cut glass appears to have insulated Libbey Glass.

The recession is described in *The Growth of the American Republic*: "Prices and wages hit rock bottom and there seemed to be no market for anything. Half a million laborers struck against conditions, which they thought intolerable, and most of the strikes were dismal failures. Ragged and hungry bands of unemployed swarmed over the countryside, the fires from their hobo camps flickering a message of warning and despair to the frightened townsfolk." The Panic of 1893 approached conditions of the Great Depression of the 1930s, although the labor unrest, socialists and anarchists made it a much more fearful time. Libbey's profits increased, however, and he was investing heavily in his development company Toledo Glass. Ohio and the nation were struggling.

Jacob Sechler Coxey of Massillon, Ohio, formed the idea of a massive march on Washington to highlight rising unemployment. Coxey owned a quarry that mined high quality silica sand for the steel and glass industries. Coxey had run for Congress as a Populist candidate. The depression had idled his operation for months as the demand for high silica sand declined. Coxey had some unique ideas that would foreshadow some in the 1930s. He hoped to form a Christian nation based on charity, and called for an eight-hour day and a minimum wage of $1.50 as well as a national bond issue to support the unemployed. Coxey suggested a massive influx of money and public works for the unemployed, but his plan for a national march struck fear in the heart of Congress. The notion of an army of tramps, unemployed people, socialists, and anarchists marching through the country struck a nerve. Many believed the country was ripe for a socialist revolution like those in Europe.

The concerns were further stirred by the press, which descended on Massillon, Ohio. While editorials denounced the march, the headlines promoted fear. Money and support started to pour in from labor unions, Populists, and probably a few Republicans. In the midst of the Coxey uproar, Governor McKinley was dealing with an outbreak of worker violence on several fronts. Coal miners from Stark County (including Massillon) were attacking trains carrying coal into Ohio from West Virginia. Rails were being torn up, trains were being shot at, and bridges were being burned. McKinley summoned John McBride, the president of the United Mine Workers, to his office. McBride noted he had "fourteen cannon pointed on one mine." His cajoling failed, and McKinley ordered a massive call up of the National Guard. The show of overwhelming force ended the violence. One train alone carried twelve men and two Gatling guns. Toledo was hit hard by railroad strikes, which also required troops. The Toledo railroad yards were a gathering point for the unemployed to supply strikebreakers across the nation. Hundreds of marchers left Toledo for Massillon.

Coxey's gathering army remained peaceful as he used torchlight parades and speeches to muster his army. Toledo sent hundreds to Massillon. Finally, Coxey gathered several thousand to start the march, with thousands who converged with his main column along the way. The body reportedly reached 20,000 at its peak in Ohio. The march started on Easter Sunday with a plan to reach Washington on May Day. Coxey and his "officers" rode horses at the front of the column, along with a band. Reporters were astonished as they rushed stories to their papers. Headlines seemed to rally more from all parts of the country. Armies from Philadelphia, Pittsburgh, and Maryland mustered as well. Coxey's Ohio army roughly followed today's Route 30 to Pittsburgh. Towns offered food as the armies passed through. As Coxey reached Pittsburgh on April 3, his numbers swelled to 300,000 in the city.[1]

Pittsburgh was the forge of the nation and the fortress of American labor. The winter of 1892–1893 had been the toughest ever for this industrial center. Hobo towns lined the great railroads leading into the city, and the streets were filled with beggars. (Even Toledo had its hobo towns around the railroad connections.) Immigrants in tent cities struggled to stay warm and eat. The rich suburbs lived in constant fear of looting or attack. The city had emptied its coffers to supply relief. Andrew Carnegie kicked in an equal amount of about $125,000, and other capitalists chipped in another $100,000. The money was used for city work projects such as Coxey had called for.

In Toledo things were similar, except for the steady smoke from the Libbey Glass plant. The army's appearance in Pittsburgh did lift the hope of the city. Bands and crowds gathered as if Coxey's group was a conquering army. But the rising hopes fell quickly as the army left Pittsburgh.

In Maryland, roads were muddy from a rainy spring and the army bogged down. Food became scarce and towns less supportive. The local authorities restricted foraging and the army dwindled with each mile. It entered Washington with under a thousand hungry marchers. Grover Cleveland had braced for trouble, and Coxey was arrested for trespassing on the Capitol lawn. Mounted police had an easy time intimidating the weary army. Some fighting broke out, but the police suppressed the army quickly. Populists in Congress read Coxey's petitions, and in the Senate, an unsuccessful bill was introduced. The New Deal in the 1930s would finally see many of Coxey's ideas implemented, and Coxey's march would be a model for the civil rights movement. His action encouraged support for a return to the McKinley tariffs, which were key to the glass industry.

For Libbey, 1895 would be monumental, with the birth and death of his only child, moving into his new West End mansion, and the formation of Toledo's Tile Club. Florence and Edward had been planning their new house for two years. The West End had been the dream of the Scott family and in the 1880s, Toledo's wealthiest homes were building there. The woods section, as it was known, was a triangle formed by the Great Trail (Detroit Avenue), the Scott property (the Toledo Museum), and Territorial Road (Collingwood Boulevard). The streets ended in "wood," such as Scottwood, Greenwood, Maplewood, and Collingwood. Collingwood was a beautiful, tree-lined avenue that was home to Toledo's richest merchants, bankers, and industrialists. It would be the first street in Toledo to be paved with asphalt.

The West End was a study in popular architecture with Queen Anne, Richardsonian Romanesque, Colonial Revival, and Italian Renaissance predominating. The Libbeys had selected self-educated Toledo architect David Stine, who had designed the Libbey showplace at the Chicago World Fair. Stine had already completed a beautiful Queen Anne mansion at 2345 Collingwood for Toledo's king of grain, cotton, and tobacco merchants, J. Frank Zahm. The thirty-room mansion was completed in 1889. Interestingly, years later, it would be purchased by Mike Owens. Stine would later be known for his design of the Lucas County Courthouse. He was said to have used a Newport, Rhode Island, house (the H.A.C. Taylor House) designed by famous architect Stanford White, which the Libbeys had visited, as a model for the Libbey House.

The Libbey mansion was an eighteen-room, 10,000-square-foot house at 2008 Scottwood. Florence's uncle, William H. Scott, was their neighbor. It was decorated on the inside with hand-carved cherry, mahogany, and oak. Woodwork included a grand cherrywood staircase, an oak-paneled dining room, and a mahogany parlor.

Ten hand-carved lion heads, each with a different expression, accented the dining room. It was a three-story home with the third floor for the servants. The stairway included a Tiffany stained glass window. One of the fireplaces was accented with Delft tiles (Mrs. Libbey was a collector of Delft pieces), and the home had an underground wine cellar. Originally the house was lighted and heated by natural gas, but later Libbey converted to electricity. Mrs. Libbey had trees transplanted from her father's nearby estate.

Mrs. Libbey had become pregnant in 1895 and bore her only son while visiting friends in Pasadena, California. Unfortunately, while the Libbeys were living at their 1825 Collingwood apartment (near Monroe Street) awaiting their new home, their only child, William Langdon Libbey, died at nine months of age on November 10, 1895. It was a difficult setback for the couple, but both dealt with it stoically. The event was rarely referred to again. Libbey was able to throw himself into his work, while Florence drew on her family and Libbey's sisters for support. The move into their new mansion certainly helped change the focus.

The couple turned more to art and travel. Florence had worked with the Toledo Art Association since its beginning in 1883. Robinson Locke (son of David Ross Locke) was a founding member of this group, which focused on art appreciation. Robinson Locke had returned to Toledo to become editor of *The Toledo Blade* after the death of his father in 1888. The Toledo Art Association had many art teachers and was involved in teaching art to local school children as well as in collecting, holding a number of exhibits throughout the year. In 1895, a somewhat secret group of male artists and supporters evolved as the Tile Club. Members included architect David Stine, animal painter Edmund Osthaus, local painter Tom Parkhurst, photographer Charles Lewis, collector Charles Ashley, and art dealer George Allen. George Washington Stevens, the future director of the Toledo Museum, was also a member. Many lesser-known national artists evolved out of the Tile Club. Florence and the Scott family were backers of this artist colony, and Libbey and Florence were early supporters of the group. Both groups would later be important in the formation of the Toledo Art Museum.

Meanwhile Richardson and the other directors of Libbey Glass continued to be compliant about Owens's use of company resources and their lack of profit-sharing in the new technology. Owens had built a wooden shop, but most testing was done in the Libbey Glass furnaces during down time. Owens told Libbey he was near a breakthrough for an automated machine to make tumblers. Libbey decided to form a new company that would focus on developing machines and processes. The split would allow Libbey to focus separately on two strategies — the branding of Libbey cut glass and the development of glass technology. The change represented Libbey's personal struggle to keep art and technology separate, but Richardson and Robinson continued to challenge Libbey's strategic decisions. Libbey was not vindictive, but he decided to keep Richardson and Robinson away from his strategic future moves, clearly remembering their opposition to his Chicago Fair project. Libbey now used Clarence Brown as a sounding board and for advice. Libbey and Brown considered Richardson and Robinson too conservative. To facilitate the split, Libbey gave Richardson and Robinson ten-year contracts to be general manager and assistant general manager of Libbey Glass. Libbey would be director and CEO of the company, but Richardson and Robinson would freely run the operations and the administration.

The development company would be called Toledo Glass. Libbey was the only shareholder from Libbey Glass. He remembered the difficulty he had in moving forward on some key Libbey Glass projects and the Owens/Richardson split could bring too many problems for a new development company. The incorporation took place on December 16, 1895. The split created two groups of glass entrepreneurs conveniently known as the "Libbey Glass Faction" (Richardson, Robinson, and Donavan) and the "Toledo Glass Faction" (Geddes, Brown, Owens), with Libbey heading both. Libbey, however, was in full control of both companies, and the schism actually increased his authority and influence in each. Libbey still had strong managers in place at both companies, allowing him more time to travel and vacation in California. Brown was a close friend and could be counted on to report by phone and telegraph on board matters. Brown was one of a handful of men close to Libbey in friendship and business.

Libbey gave Owens enough shares to make him the second largest shareholder of Libbey Glass. He was followed by James Brown Bell of the investment firm of Secor & Bell with 246 shares. Others included Clarence Brown (Libbey's lawyer and friend), Frederick Geddes (from the same law firm), Samuel Stump, William Tyler, and William Harrison. Tyler and Harrison were also lawyers at the firm of Geddes and Brown. Most of these directors had but a few shares and were more advisory in nature initially. Even with his shares, Owens was not elected to the board. He had no interest in being a director, and Libbey actually didn't want him on the board. For Owens, freedom from close supervision was enough. Libbey's old friends at Libbey Glass were clearly cut out. Libbey, however, would still make requests and draw on the resources of Libbey Glass when needed. He did draw back from the day-to-day operations of Libbey Glass, evolving into a true CEO and chairman of the board. At the first board meeting, Libbey and Owens were voted 375 shares at a value of $37,500 for patents and work done prior to the company's formation. Dean V.R. Manley, president of National Cement and Home Saving Bank and treasurer of the Toledo Club, was brought in as an outside director.

Toledo Glass was doing development work on a bottle machine and selling rights for the Owens semi-automatic system. The company was facing a number of competing patents, and Libbey decided to purchase the patents instead of suing. The

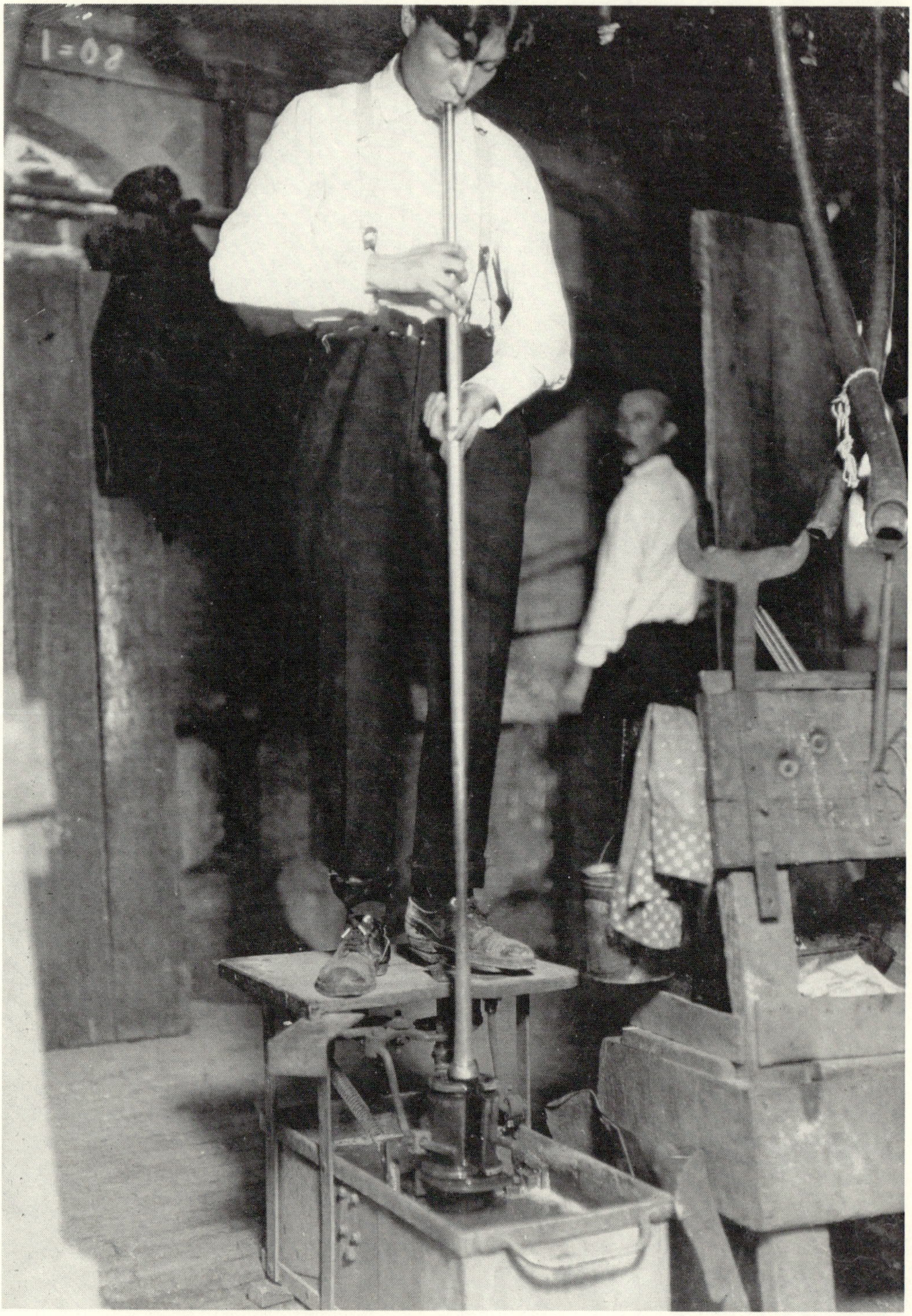

Bottle blowing, 1897 (courtesy of University of Toledo, Ward Canaday Center, Libbey Glass Records).

semi-automatic process eliminated skilled blowers and reduced operating costs. A semi-automatic machine with a three-man crew could do 4,300 bottles or jars a day, about a 20 percent increase in production. Widespread acceptance of the machines meant that Toledo Glass had to go a step further in automation to stay ahead. Competing semi-automatic systems were developed by people such as Philip Arbogast of Pittsburgh. It is clear in hindsight that development work in the industry was moving toward full automation. Libbey was ahead with his focused development company. The semi-automatic process was changing the union and industry by 1903. With skilled blowers eliminated, gatherers became the masters of the work hierarchy. Management successfully changed the wage structure to incentives for increased production. The unions were also having jurisdiction fights as the industry changed. In the container industry in general, the Glass Bottle Blowers' Association was in competition with the American Flints. The American Flints fought hard to keep wages high and limit production in semi-automatic plants.

Few understand that Owens, while an inventive genius, required extensive engineering and mechanical support. One of the first investments of Toledo Glass was in mechanics and draftsmen. Owens lacked the ability to produce the two-dimensional drawings required to manufacture parts. One insider recalled: "One of Owens's strange attributes was, in spite of his inventive genius, his inability to visualize things. He wasn't mechanical."[2] In those days, engineers made two-dimensional drawings, and Owens produced very crude three-dimensional ones, as did Edison, who visualized in three dimensions. Owens draws a lot of attention, but it's more productive to consider Libbey's role. Libbey created a supportive environment for Owens's genius. Libbey saw Owens as a motivated project manager, capable of results. Owens was the Napoleon of project and developmental work. Libbey supplied the resources that Owens needed. Toledo Glass was every bit Libbey's organization. The engineers and mechanics hired by Libbey were also skilled with patents. Owens added his brother, Thomas, to the organization after he complained that all these "engineers" were hurting his progress.

Toledo Glass would be novel as a plant to develop processes and equipment in a real world environment. This was a step beyond Edison's experimental Menlo Park and George Westinghouse's experimental building. The idea was Libbey's. To some degree, it was to isolate Owens from causing more tension at Libbey Glass, yet it was a new approach to research and development. The plant was built on the corner of Detroit and Delaware avenues in the newly annexed Auburndale area. It would have a fourteen-pot furnace designed by Mike Owens for tumbler development, along with building for engineering and drafting. Mechanics and draftsmen were hired immediately to get things going, and some glassmakers and blacksmiths were transferred from Libbey Glass. The first two key employees were engineer Emil Bock and mechanic Henry J. Colburn. Owens had suggested they start work on a glass blank

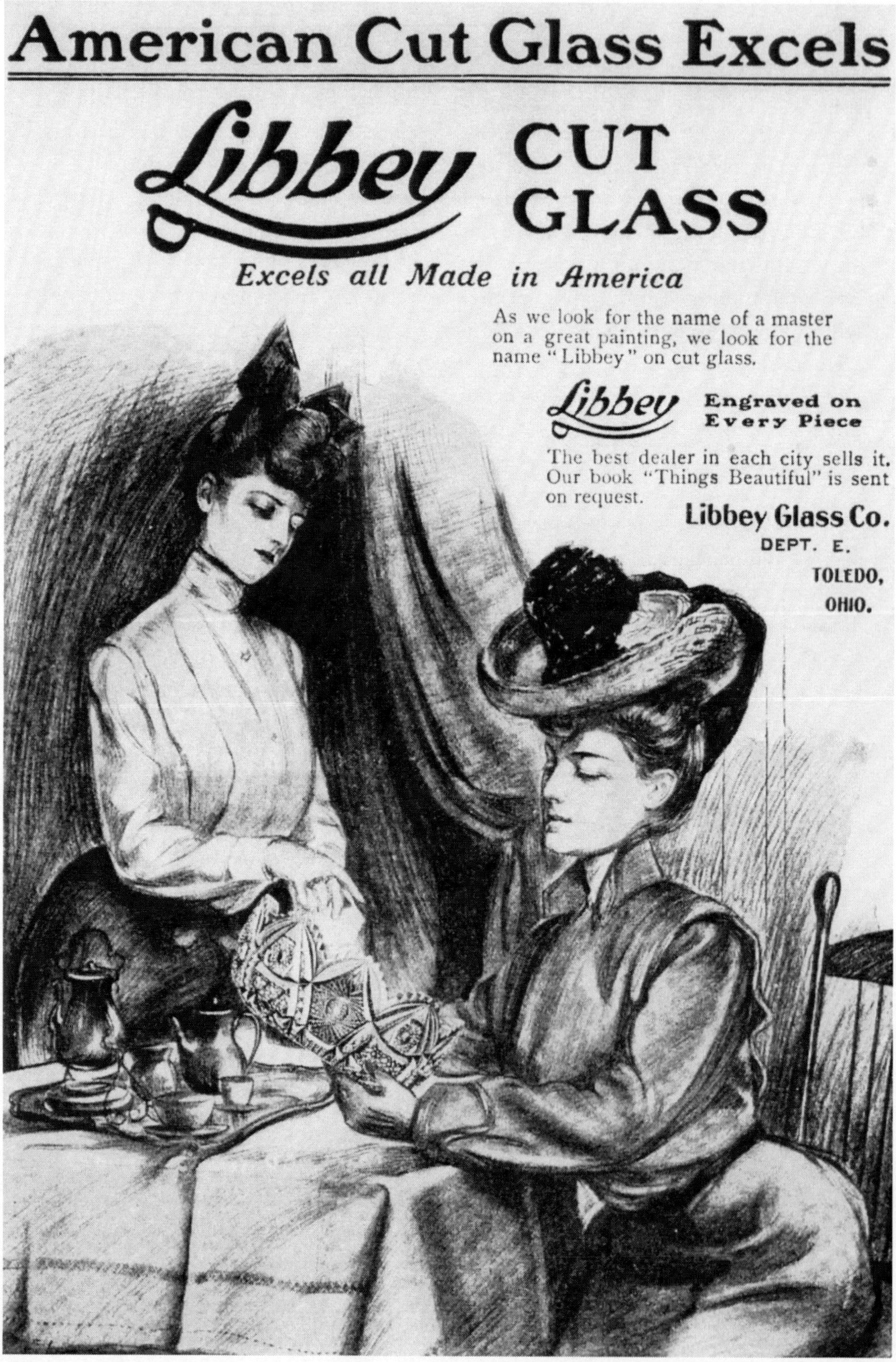
American Cut Glass Excels
Libbey CUT GLASS
Excels all Made in America
As we look for the name of a master on a great painting, we look for the name "Libbey" on cut glass.
Libbey Engraved on Every Piece
The best dealer in each city sells it. Our book "Things Beautiful" is sent on request.
Libbey Glass Co.
DEPT. E.
TOLEDO, OHIO.

machine to sell to Libbey Glass, but Richardson said he was not interested in working with Owens. Richardson's decision only widened the schism. Libbey had recently added a new manager to Libbey Glass, his brother-in-law, William S. Walbridge. Walbridge's furniture business had failed during the Panic of 1893. Once Toledo Glass was fully operational, Walbridge moved over to Toledo Glass to manage the finances. It would be a challenging job, because Owens only knew how to spend. Like Clarence Brown, Walbridge had Libbey's total confidence in all matters. Walbridge, Brown, Geddes, and Owens would be the core of Libbey's glass empire. They were extremely loyal to Libbey and in return would prosper with his success. They all would become multi-millionaires and would serve on interconnecting boards of various glass companies.

Opposite: **Libbey Glass ad, 1900 (courtesy of University of Toledo, Ward Canaday Center, Libbey Glass Records).**

Chapter 10

The Toledo Glass Company

Toledo Glass would be the core of Libbey's future business, and the engine of his success, America's first corporate research and development company. With a stable market and future as a result of the McKinley tariffs, Libbey invested heavily in the future of American glass. Toledo Glass worked on a wide range of development projects in many segments of the glass industry, including electrical automation of older processes and equipment. Electrical controls replaced hand and foot levers. With the formation of Toledo Glass, Libbey Glass and its management only produced tableware, bulbs, and cut-glass. This would create a tension between the two companies, although agreements had been made with Richardson to protect Libbey Glass from any deals by Toledo Glass that would be injurious. Most of these agreements were verbal, since Libbey was the major shareholder and force behind both companies. Libbey acted many times on his own without conferring with the board of either company.

At Libbey Glass, his actions were resented, and bitterness grew between Richardson and Libbey. Mike Owens continued to be a problem, crossing company lines at will. Libbey Glass needed further development in bulb making and tableware manufacture, but Owens was interested in moving beyond automatic bulb-making. Libbey Glass had purchased the domestic rights to the Owens bulb-making machine, but it was clear that Owens's work could be used to make a wide variety of products. Libbey Glass and Richardson wanted more development work on the bulb process. After bulbs, Libbey focused on developing the Owens semi-automatic process for tumblers. Libbey Glass could benefit but Owens had a bigger view, looking to possible bottle manufacture.

Tumblers for taverns and the home were big business and were not part of the Libbey Glass product line at the time. The shape of the tumbler was perfect for an adaptation of Owens's process, and Libbey agreed with the strategic shift to the development of tumblers. Owens had a tumbler paste-mold system ready in 1896. Libbey handled the sales, approaching the Rochester Tumbler Company, the largest tumbler manufacturer in the world at 45,000 dozen tumblers a week. The plant took up seven acres and a half-mile of riverfront on the Ohio west of Pittsburgh. The plant had over 1,200 employees producing 75,000 tumblers per week. The Owens paste-mold system could produce 200,000 tumblers and reduce costs by 40 percent. This was Libbey's first crack at selling machines, and it would be a learning experience. The agreement called for Libbey to supply six machines. In return, Rochester Tumbler would pay $100,000 over a ten-year period in yearly installments. Installments were necessary because of the still-poor economy in 1896. Rochester would flood the market with tumblers, stemware, finger bowls, lemonade glasses and similar products. Once again, Libbey Glass management felt short-changed.

The year 1896 would be critical for Libbey and his companies as well as the glass industry in general. Ohio Governor William McKinley was running for president. Libbey was a friend and a financial supporter of the McKinley campaign, which promised to restore the manufacturing tariffs to their 1890 levels. Over the year, McKinley had been a frequent visitor to the Toledo Club and had developed strong financial support in the Toledo area. The poor economy favored a return of the Republicans to power and McKinley won big. With his 7,101,401 votes and 271 electoral votes to William Jennings Bryant's 6,470,656 popular votes and 176 electoral votes, it was the biggest Republican victory since Grant. The victory would augur boom years for America and the glass industry. The improved outlook made money more available for development investment and capital spending. In 1897, McKinley was able to return the tariffs to 1890 levels, a favorable condition for the glass industry. Tariff levels ran from 30 percent to 45 percent on the value of the shipped product.

Libbey saw the growth coming and personally shifted his focus to Toledo Glass, where he could profit the most. But the glass boom was creating record profits and dividends at Libbey Glass as well. Libbey Glass was employing over 900 workers and over 150 glasscutters. The high-end market for cut glass was particularly profitable under the McKinley tariffs. Libbey and Tiffany collaborated on cut glass and sterling silver pieces. These products made Libbey extremely proud. He was exporting many of these high-end products, having a cost advantage by using automated processes.

Libbey went to England in 1897 to sell European rights to the tumbler machine. Prior to the visit, Toledo Glass had applied for patents in Russia, England, Spain, Italy, Austria, France, Belgium, Norway and Sweden. Furthermore, Toledo Glass had hired Samuel Snell as an agent, in the hope that the patent rights or machine rights could be sold readily in Europe. Libbey's 1897 European trip, like so many before and after, mixed business with endless visits to museums and historical sites. The initial visits with glassmakers proved that there was little interest in semi-automatic glassmaking, but Libbey would leave Snell to look for further opportunities. The guild system, socialists, cheap labor, and even government opposition to automation were major roadblocks in Europe. In addition, there were competing semi-automatic systems in Europe. While these were similar to Owens's machine, the patent rights were older. This provided further motivation to proceed with work on a fully automated process. Labor in Europe cost about half what it did in America, so European manufacturers valued automation less. Europe was developing a line of less capital-intensive semi-automatic processes.

The major competing semi-automatic processes were those of Pittsburgh's Philip Arbogast in America and of Howard Ashley in Europe. Arbogast used a "press and blow" process, patented in 1881 that was popular in the Pittsburgh area. The process first pressed a blank in a mold and then transferred it to a blowing mold for its final shape. Originally it was used to produce Vaseline-type jars. The Arbogast process was particularly suited for wide-mouth jars and thick-walled jars. Interestingly, the union reduced wages of blowers and workers in an effort to cooperate with the semi-automatic process. The more labor-intensive Arbogast process was preferred to the Owens process. Libbey and Toledo Glass initially decided to focus developmental work on the thinned-wall market of chimneys, bulbs, and tumblers, but Libbey watched closely the success of thick-walled jars in the growing fruit and vegetable packaging market.

The H.J. Heinz Company, a major user of jars and wide-mouth bottles, was using the Arbogast process. Heinz had been an innovator in glass packaging as early as the 1870s. Glass bottles were the key to Heinz's success; he realized that the package was as important as the product. Although he started his business at home, Heinz used the best processing science and packaging available. Using glass jars, he applied the preservation techniques of Nicholas Appert, who had published the *Book for all Households, or the Art of Preserving Animal and Vegetable Substances for Many Years* in 1810. Accordingly, horseradish bottles were immersed in boiling salt water for about thirty minutes. Heinz's product and packaging were better than most of the imported and commercial domestic products available. Packaging had always been a unique part of Heinz's business. Living in the "Glass City" of Pittsburgh, he loved to use special bottle shapes as well as clear glass. Heinz had pioneered the use of glass table decanters with glass lids. As business volume increased under Heinz, Noble & Company, Heinz designed his own glass mold for his bottles and glass decanters. The table decanters had an embossed anchor on the side. More basic bottles had Heinz, Noble & Company embossed in the mold. Heinz loved to visit local glass houses to get ideas on packaging and was the first industrialist to enter the food processing and distribution industry.

Heinz became the first domestic company to produce bottled malt vinegar, which was popular with the consumer. His bottling of table vinegar was a marketing innovation for the time. Vinegar was usually sold to the grocer in barrels and then resold in pails. Bucket sizes varied, making pricing difficult for the grocer, and bucket handling was difficult for the consumer. The same was true for pickles and sauerkraut. This moving from barrels to bottles was another key product strategy which allowed direct sales to grocers, and Heinz promoted this strategy throughout his career. Bottles were about 2 to 3 cents; while not a major cost, it was significant. Prior to 1890, fancy bottles cost from 30 cents to 60 cents a bottle, which made packaging the major cost in bottled food. Semi-automatic hand blowing methods brought the cost down from 5 cents to 10 cents a bottle by 1895. At the turn of the century, glass bottles used in high volume such as Heinz products used reached $2 a gross (144 bottles). Heinz supplied his own bottles for ketchup and surpassed the million mark. The best glass house at the time used hand blowing and could produce 3,600 ketchup bottles per furnace operation in a day.

Heinz had quickly become involved in the design of the bottles, and in the 1880s would patent various shapes. In 1890, Heinz patented his famous octagon ketchup bottle to prevent the use of his private molds by his food processing competitors. The glass bottle had always been key to Heinz's sales strategy, and he had patented a number of designs in 1882 for further protection. By 1890 Heinz was producing a third of his glass bottle needs. As for the glass molds, Heinz took control of them to prevent copying. He formed closer partnerships with glassmakers and took some business to Wheeling, West Virginia, glassmakers to further prevent copying by Pittsburgh glassmakers. West Virginia glassmakers were also advancing new bottle technology such as screw top lids. One of those West Virginia companies, Hazel-Atlas Glass, was part of Heinz's supply chain. Hazel-Atlas was one of the largest and most efficient bottlemakers of the time. D.C. Ripley of Pittsburgh was another of his suppliers, who used the Arbogast process. It was clear that without an automatic machine, Libbey would not be able to sell to the Heinz Company. Another company using the Arbogast process was Ball Brothers Glass Manufacturing Company of Muncie, Indiana. The home canning industry, which the Ball Brothers controlled, had become the hottest growth segment in glass in the 1890s. Libbey and Owens were closely watching the glass bottling boom. Libbey would have to make a major advance in automation to pursue that market. Libbey, in the short run, looked for more immediate markets for the Owens semi-automatic process.

For the time being, Toledo Glass under Owens continued to produce thin-walled tumblers, as he developed a paste system to manufacture kerosene lamp chimneys. The improved thin-wall process allowed Owens to look to new markets, particularly the glass chimney business. Owens wanted to develop products, while Libbey still favored selling machine rights. Libbey was feeling too stretched in his focus, and his only real production interests were at Libbey Glass. Owens, however, had some experience in blowing chimneys. Libbey and Clarence Brown came to realize the potential of taking over this segment of the glass market. Libbey had assigned James Brown Bell and Clarence Brown to explore the potential of the chimney market.[1] They estimated that hand-blown chimneys cost 23.13 cents per dozen, while Owens's paste mold could produce them at 10 cents per dozen.[2] The thin-walled chimneys were all hand blown in the Pittsburgh area. The paste-mold system cut out the blower and gatherer, the top labor costs. The production rates were significantly higher, putting Libbey in a position to control the entire market and make a huge profit competing against hand-blown manufacturers. Libbey had missed the opportunity to do a similar thing in the tumbler business. The McKinley administration would look favorably on such segment control, so Libbey moved forward.

In February of 1898, the Toledo Glass shareholders were: Edward Libbey, 642 shares; Mike Owens, 275 shares; James B. Bell, 280 shares; Dean V.R. Manley, 95 shares; N.A. Goodman, 50 shares; H.H. Nood, 55 shares; Clarence Brown, 51 shares; Fredrick Geddes, 51 shares; and the firm of Brown & Geddes, 1 share.

A new company incorporated in April 1898 as the American Lamp Chimney Company. Libbey created the company through Toledo Glass by issuing stock of American Lamp to Toledo Glass. In effect, Toledo Glass owned American Lamp and the investors got back their investment in Toledo Glass in cash. The American Lamp Chimney Board reflected that ownership with its directors: Libbey, Clarence Brown, Fredrick Geddes, William Walbridge, James Bell, and Dean V.R. Manley (a Libbey friend and local banker). Libbey controlled a little over 50 percent of the company. Libbey added some other business associates, including West End neighbor and spice king William Brigham, lumber king Jetro Mitchell, Charles Gardner, and James Hutton (a New York investor). Libbey was named president and his brother-in-law, W.S. Walbridge, treasurer. Mike Owens was named plant superintendent and Walbridge his assistant. Libbey had no choice but to give Owens the position, but he and the board worried about Owens's lack of financial management skills. The new company not only took over Toledo Glass's equipment and plant, but expanded the structure.

American Lamp was a direct threat to the companies of George Macbeth and Thomas Evans. Both of these producers controlled a market of over 25 million lamps per year. Both companies were Pittsburgh-based with plants in Pennsylvania and Indiana. Macbeth had just built a state-of-the-art chimney plant in Marion, Ohio, which had a capacity of 12 million chimneys per year. They were hand-blown operations, which would doom them in head-to-head competition with the Owens process. Evans and Macbeth went directly to Edward Libbey to discuss a solution. Libbey discussed the possible options with James Bell and Clarence Brown, both of whom preferred to sell and lease technology than to manufacture products. They agreed to sell the Toledo Glass chimney operation for cash and stock in a new company, Macbeth-Evans. Shareholders in Toledo Glass received a 100 percent dividend with shares in Macbeth-Evans. Toledo Glass would control about 15 percent of Macbeth-Evans. Libbey had once again turned development and research investment into cash. Libbey served as a vice-president and director of the new company. Mike Owens was superintendent of the Macbeth-Evans Toledo plant.

This did not slow Michael Owens, who had been working on other glassmaking processes. Another Owens invention in 1899 would finalize the growing schism between Libbey Glass and Toledo Glass. Libbey Glass had a major piece of business in "glass blanks."[3] The blank business allowed Libbey to compete to produce a cheaper and more profitable tableware piece than cut glass. A blank could be blown or pressed in an engraved mold, then rough cut to sharpen the lines using semi-skilled labor. Owens perfected the blowing of blanks into molds. Still, the blank had an internal "fogging" that was not easily corrected. Owens invented a fire polishing process to make quality blanks. He then offered to sell his invention to Richardson at Libbey for $1,500. Richardson rebuffed Owens; while he wanted the invention, he did not want to pay for it. Richardson's point was that Libbey Glass had been involved in the development of the process. Richardson, of course, had no love for Owens, but Owens would turn out to be more than fair. The move was consistent with how Libbey had often done business — by selling to his own companies. There is good evidence that both Owens and his brother Tom often tested processes at Libbey Glass on the back shifts. Owens also worked with C.J. Nolan of Libbey Glass. Owens drew no corporate lines between Libbey's companies.

Owens went to Fry Company in Pennsylvania and negotiated a deal for $5,000. Fry Company would have exclusive rights to the process under the deal. Owens wired Libbey in California and got permission. It would be a costly mistake and one that ultimately produced a lawsuit between his own companies. Libbey was on vacation and probably did not fully consider the ramifications of such a deal. He always had trouble saying no to Owens and preferred to deal with Owens through other officers so projects could be fully vetted. Many saw Owens as the son Libbey never had. Clearly, Libbey had a weakness for requests by Owens. It had been the reason he wanted Clarence Brown to be president of Toledo Glass. The process turned out to be very successful, producing a highly competitive product. Libbey Glass lost significant business and had to purchase a non-infringing patent from C.J. Nolan to compete. Over the years, Libbey Glass lost millions of dollars to the Owens deal, and Libbey was forced to clean up the mess between his two companies. He suggested Owens be made a consultant to Libbey Glass, which would legally prevent him from making such damaging arrangements. Libbey Glass did this, but it opened their operations again to Owens for experimentation. Libbey also negotiated an agreement with Fry Company over the blank business. For his part, Libbey started to pull away from his easygoing relationship with Owens.

In 1897 and 1898, Toledo was producing thin-walled tumblers and had employed over 250 workers. Mike Owens had started a project for automated bottle-making. During 1898, Owens perfected a single "hand gun" bottle-making device and applied for a patent on December 26, 1899. Really the device could only make thicker-walled and wide-mouthed small bottles and jars. This first step towards the automatic bottle-making machine was a confluence of ideas, inventions, and men, but Libbey was the channel that brought them together. Owens played the central role, but it was more complex than the actions of a single man. All of Owens's semi-automatic machines used compressed air blown through a pipe, simulating the hand blowing of the glass blower. His gun device was based on a piston for sucking molten glass. The gun device roots stem from a sucker-upper device used as early as 1890.[4] The sucker-upper was a cup at the end of a piston pump. It was placed in a pot of molten glass, and by pulling the piston and creating a vacuum, a cup of glass was gathered. The gather could then transfer it to a rod or blowpipe. As a worker perfected it in practice, the quantity gathered was standardized. The device was never patented. There is some evidence that Mike Owens was experimenting with it at the World Fair in 1893.

Owens's bottle-making gun was nothing more than a specialized sucker-upper, often described as a bicycle pump. During the same time period, a number of innovators were working on a similar device throughout the world. But even this crude device was the work internally of Owens and Emil Bock, whom Libbey had hired to help with the mechanical aspects. Bock would prove to be a great engineer, and he played a key role in the ultimate development of the automatic bottle machine. Charles Schwenzfeier and Seth Hanshaw were important mechanics in the building of devices. Libbey often used Clarence Brown to keep Owens from firing these men, who often clashed with him. But Owens's spending on the project in the early going was more problematic. Libbey had his brother-in-law Walbridge working closely with Owens on spending. Owens could, however, be quite charming and persuasive. In 1899, the bottle machine became the main project of Toledo Glass, and some years of deficits followed. The cost to get to an operable machine in 1904 would be over $500,000. Libbey was well aware of the booming market for bottles. Ketchup and milk bottles alone meant five million bottles a year. Jars and bottles were being used in all forms of packaging. Home canning was becoming popular by the late 1890s, as the price of glass containers brought it within a family budget. The Arbogast process was used for Mason and fruit jars, but Libbey saw a way into the market through technological advances.

In October of 1899, the Toledo Glass Board authorized hiring Emil Bock as Owens's assistant for $40 and a stock bonus if the project was successful. Owens would receive $2,000 annually. He was also given funds to hire mechanics to back up the project. While the automatic bottle machine would be a breakthrough invention, it almost broke Toledo Glass financially and internally. Still, Libbey saw it as necessary to break into the bottle and fruit jar markets, which were huge and growing. Libbey allowed the bottle machine to advance. Owens was extremely difficult to work with, and Bock and others often disagreed with Owens. Just as problematic was Owens's approach to board members. Clarence Brown had clearly lost patience with Owens by 1900. Owens's main ally became Walbridge, who took a personal liking to him and was not close to many of the board members. Walbridge never fully embraced Toledo, maintaining a summer home in Pike, New Hampshire, where the Libbeys often joined him and his wife for summer weeks. Libbey had to hold the whole thing together. Libbey's own personal project for the period was not the bottle machine, but the building of an art museum. It was during this time that on a short trip to Florida, Libbey took up the game of golf. On his return to Toledo, he persuaded the directors of the Toledo Country Club to add a golf course. Libbey donated $500 and some land to have the course built.

Chapter 11

Signature Years

The signature achievements in Libbey's life were the invention of the automatic bottle-making machine and the formation of the Toledo Museum of Art. Libbey had formed the world's first corporate research and development company in Toledo Glass. He was both a glassmaker and a seller of technology. His cut glass made at Libbey Glass Company remained world renowned. Libbey Glass was a premier tableware manufacturer. He had entered and was dominating the electric bulb market and oil chimney market. He created markets in glass blanks and tumblers as well. With his Toledo operations, Libbey had taken the title of "glass capital' from Pittsburgh and given it to Toledo. Toledo was glass, and Libbey looked to give this industrial town an art museum. More importantly, he had a vision of art, industry, and community in Toledo. He would apply his successful industrial management system to the development of a community museum that would bring him world acclaim.

Glassmaking had hardly changed in three thousand years, until the invention of the bottle-making machine. The automatic bottle-making machine would be a true manufacturing revolution, comparable to Eli Whitney's cotton gin in the cotton industry. Even more, the bottle machine changed society and culture. A fifteenth-century Venetian would have felt right at home in an 1870 glass house. But the machine led to a revolution in packaging, with milk bottles, beer bottles, pop bottles, baby bottles, and glass jars. Business friends Adolphus Bush and H.J. Heinz were pioneering glass packaging in beer and vegetables but lacked volume suppliers. Early machines produced bottles at about 10 cents a gross versus the $1.52 per gross cost for hand production. The semi-automatic process could make bottles at 75 cents a gross. Lower costs quickly opened up new markets for glass bottles and jars. The automated bottle machine led to automated filling machines for foods such as ketchup. The bottle machine guaranteed standard weights and measures, paving the way for the Pure Food and Drug Act of 1906. An ancillary effect was the eventual elimination of child labor in the glass industry. But just as important for Libbey personally was the creation of an art museum.

While Libbey's Toledo Glass was the developer of the automated bottle machine,

Libbey's role is often downplayed. Michael Owens gets the credit (rightfully so for the invention), but full development and implementation required the backing of Toledo Glass, and Libbey gets credit for that. Libbey built Toledo Glass to do development work, and without it, Mike Owens could not have made the bottle-making machine commercial. Libbey hired the draftsmen, mechanics, and engineers needed to build the machine, which was an organizational triumph. Toledo Glass engineer Richard LaFrance would say: "If you locked Mike Owens inside a machine shop in 1890 and let him out in 1920 he wouldn't have made a thing. If you had done the same thing with [Emil] Bock, he might have come out with a machine, but it wouldn't have made bottles." The bottle machine required an army of technical experts and Libbey's leadership.

The bottle machine had nearly 10,000 individual parts, while the 1907 Model T had fewer than 3,000 parts. To be designed and manufactured, every part required a two-dimensional blueprint. Amazingly, Mike Owens could not read blueprints. Multiple machines meant standardized parts, blueprints, and processes, which were beyond Owens's abilities. It was a huge undertaking, and even more difficult when your top managers had huge egos, such as Emil Bock, Mike Owens, and Solon Richardson. Libbey's organization and Libbey the CEO facilitated the invention of the bottle machine.

Libbey not only held the organization together, but also funded the vision when the board of directors wavered. The bottle machine was one of the first real team inventions. It marked the end of the lone Victorian scientist making a solo breakthrough. Edison had an organization but was mostly a one-man show, while Libbey had a true R and D organization. Libbey approached research and development as a craft, with engineers as innovation artisans. Neither America nor the world had ever seen a company formed for the express purpose of innovation. Libbey drew on his history as an innovator at his father's firm to set Toledo Glass on the road to development. He had learned to bring skilled people together at New England Glass. More importantly, Libbey found a way to make innovation pay. Toledo Glass would become the American model for research and development centers of the future.

With the start of 1900, Owens and his Toledo Glass team were working full time on a fully automatic glass bottle machine. Owens's bicycle pump idea worked, but he needed a practical application. Emil Bock had drawn up a set of blueprints, but Owens had no use for blueprints and wanted it in iron. Bock and C.W. Schwenzfeier built an experimental machine based on Owens's oral instructions, rough notes, and crude drawings on a blackboard. It was a two-part mold on a vacuum pump device. Most of this work was done at a shed on Ash Street. This early experimental machine only produced a perfect bottle after several attempts. Frustration often led to arguments between Owens and Bock. Bock was methodical in his engineering approach. It drove Owens crazy, but it would be the key to the manufacture of glassmaking machines, which were not simple mold release systems that could be made in blacksmith shops. Owens complained to Libbey, but Libbey wisely stayed out of it. Often Owens would have his brother, Tom, working secretly on some part of the project. Owens and his brother might also work at Libbey Glass during the back shifts, which drove Solon Richardson crazy. Walbridge played the role of mediator and kept Libbey up to date on the project. They didn't realize that they were working on the most complex machine ever made.

The frustration of the Toledo Glass board can be seen in its December 26, 1900, report to shareholders: "Progress upon the machines upon which William Emil Bock has been working is far satisfactory that this company desires said work to continue."[1] Owens was not mentioned, although the board tended to refer to the project as the Owens/Bock project. Furthermore, Owens had no title while, Bock was given the title of superintendent. Owens, of course, was a director, although he rarely attended meetings. He was constantly at odds with the board, and he often let them know he thought little of lawyers in the glass business. The organization often reflected the sense of confusion about who was in charge. Libbey realized that both men were needed, and as at other times, tended to let things work themselves out. Libbey refrained from taking sides or changing the organization. Like Andrew Carnegie, he had learned to manage disagreement. Libbey allowed the board to pressure Owens until Owens would come to him and complain. The board would very respectfully disagree, but the board always deferred to Libbey. In most business organizations, the tortured relationships and dysfunctional organizational structure would not have been sustainable; but Libbey knew genius and creativity required an organization to accept such tension. He had dealt with the egos of glass artists for years.

In the winter of 1900–1901, Owens became bedridden with a cold. The project gained some ground, with Owens out of the picture for weeks. Emil Bock was an outstanding engineer and well versed in machine operation and machine design. Bock had the respect of Walbridge and the Toledo Glass board. Walbridge and Bock hired a Detroit machine shop to make a better model in the spring of 1901. This second machine was built on wheels. It required an operator and several helpers to produce a small, inkwell-size bottle (a petroleum jelly bottle). The experiments went on through 1902 with limited success. Costs were mounting and so was Libbey's frustration with the project. The board was convinced that Owens was the problem, with his fiery personality and blacksmith approach to machine design. Libbey realized he needed plans so machine shops could reproduce the design. Walbridge and Libbey hired an assistant for Emil Bock, Richard LaFrance, and set up a drafting room in Toledo's Spitzer Building. In early 1902, Bock and LaFrance's blueprints resulted in a third prototype, which was a big step forward. This was a five-mold machine that was motor driven, but its operation proved problematic. The machine dipped molds in molten glass and sucked the glass up into the mold using a suction pump. The constant dipping of cast molds in one spot cooled the glass, rendering it less fluid. The cooling also limited the ability to produce longneck bottles.

Bock's mind seemed blocked. Owens, however, would make another breakthrough. The chilling problem had no simple engineering solution. It required what we call today thinking outside the box, and this was Owens's specialty. Both Libbey and Walbridge did everything possible to encourage him. Costs were going through the ceiling, and Libbey was now putting personal money into the project. Toledo Glass had not paid a dividend in two years, and everybody's patience was thin. Clarence Brown wanted to put Bock in charge of the project, and all of the board agreed except Libbey. He had faith in the organization with Owens at the helm. Owens was said to have been frustrated to the point of quitting. Libbey held the project together in the difficult year of 1902, with cash from continuing record profits each year from Libbey Glass. Owens finally came up with the idea of a revolving pot to overcome the stationary dipping problem and its inherent glass cooling. Implementation, however, proved difficult. At one point when a Bock-built machine failed, Owens allegedly attacked the machine with a sledgehammer, yelling to the "damn Dutchman" to make it right.

This re-work resulted in a fourth machine, which was ready in mid–January 1903. The new machine could make longneck bottles, something that had eluded earlier designs. The final design had cost an additional $500,000, much of which came from Libbey's personal funds. The final trials included the full team of Mike Owens, Emil Bock, Richard LaFrance, and machinists Jim Cunningham, Win Rohr, Harry Brobst, Tom Feeney, Tom Mitchell, Lester Dull, Mac Linn, John Martinek, and Harry Spanier. Libbey and Walbridge were also there. This successful design became known as the "machine of a thousand gears." Actually there were 9,576 parts, making up the world's most complex machine. The first bottle it produced was a beer bottle. According to W.S. Walbridge, he took it to Toledo's Finlay Brewing and had it filled with beer, to be opened when "the firm pays dividends."[2] The bottle was never opened and was last reported in 1934 in Walbridge's private bottle collection. However, Libbey and Walbridge negotiated a limited deal to make test beer bottles for Toledo's Finlay Brewing.

The machine's statistics were impressive. It could produce eight pint beer bottles a minute. The labor cost was estimated at 8 to 10 cents a gross versus $1.80 for hand production.[3] A regular bottle-blowing crew or "shop" consisted of a blower and assistant, a gatherer and assistant, and a finisher and assistant. The crew also had a mold boy, snapper-up boy, carrying-out boy and carrying-in boy, plus shared cleaning boys. The machine eliminated the blower, gatherer, and finisher with assistants — all skilled positions. The mold boy and snapper boy were eliminated. The Owens machine still required a machine operator and helper, and boys for the annealing process. The real savings of the machine was in the increased productivity. The commercial version in

1905 known as machine A could produce twelve beer bottles per minute, or 17,280 in a 24-hour period. This compared to hand production of 2,880 per day using a crew of six men and boys.[4] Such production capacity and lower costs quickly opened up markets for beer and ketchup bottles. The national market for ketchup alone in 1900 was about three million bottles. Since ketchup was sold in barrels as well, the potential market was probably near seven million bottles. The beer bottle market in 1900 was about twice that size. Libbey fully realized the potential.

The editor of the glass industry journal *National Glass Budget*, Frank Gessner, summed up the bottle machine:

> It gathers its glass, forms its blanks, transfers the blank from the gathering mold to the blow mold with a finished lip and ring, blows the bottle and delivers the finished bottle automatically without the touch of the human hand, eliminates all skill and labor and puts the same amount of glass into every bottle, makes every bottle of the same length, finish, weight, and capacity, it wastes no glass, uses no pipes, snaps, finishing tools, glory holes, rosin, charcoal, and requires neither gatherer, mold boy, snap boy, nor finisher and still makes better bottles, more of them than any other process.[5]

The industry and union braced for a true revolution; their future was now in the hands of Edward Libbey. The Owens machine would reduce the glassmaker to a machine operator.

Still, the prototype was not ready for commercial production just yet. The machine was moved into a new demonstration plant on Libbey Street. Emil Bock's contributions toward commercialization of the machine were critical. Bock standardized the parts, using gauges and fixtures. Standardization would allow for easy production of the machines. Blueprints could be given to machine parts manufacturers. The principle of parts standardization and the use of fixtures and jigs was ahead of its time, and enough for Libbey to launch a sales campaign. He lined up Pittsburgh Brewing and Adolphus Busch for a Toledo demonstration. H.J. Heinz also went to Toledo to view the machine and consider its use in making ketchup bottles, which might have been the biggest bottle market. Libbey planned to advertise the machine at the St. Louis Exposition in 1904. In late August, the board of Toledo Glass moved to open a new company to sell the bottle machine. On September 3, 1903, Owens Bottle Machine Company was formed. It had the same directors at this point as Toledo Glass, but Mike Owens had a larger share. Finally, Libbey could relax a bit from the last four years of development work and turn to other interests. The year 1904 would usher in new business and personal projects. Libbey would turn his passion for development to creating an art museum.

Libbey's love of art had deep roots. Clearly, he always considered glassmaking to be a true art. In his first trip to Europe, he studied with great glass artists. His second trip to Europe was a grand tour of museums and architectural sights, which he detailed in his travel diary. He first learned the broader forms of art collecting

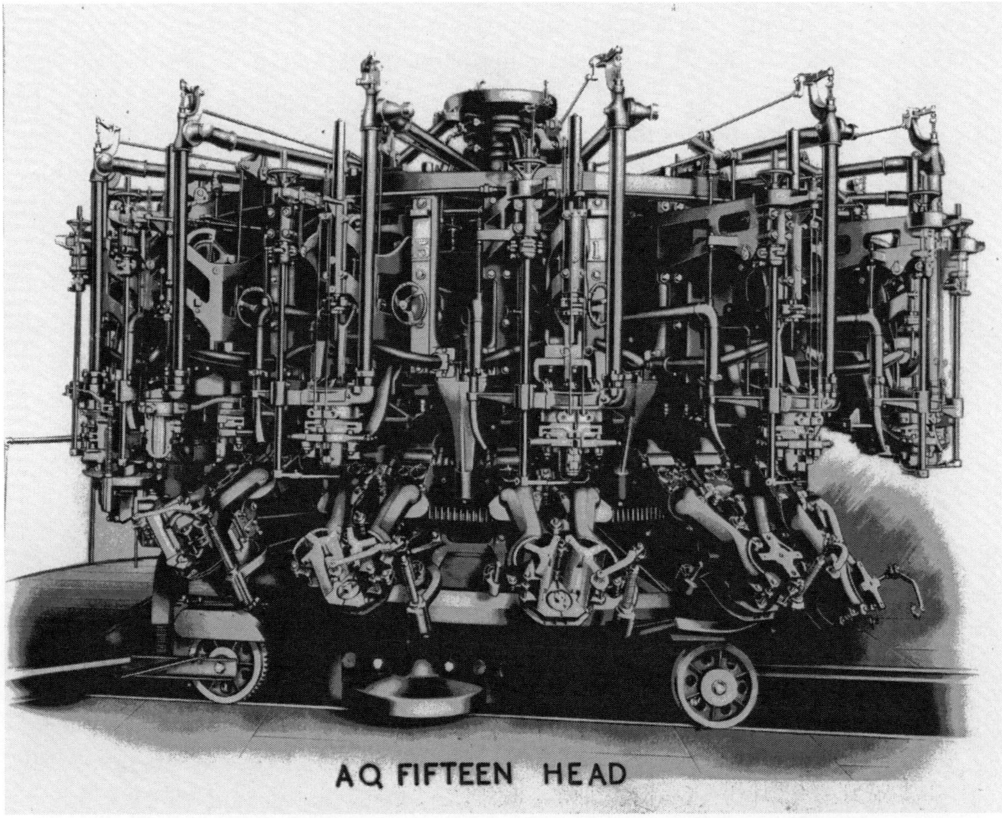

Early, developmental Owens machine. (courtesy of University of Toledo, Ward Canaday Center, Libbey Glass Records).

from the son of his father's partner, James Jackson Jarves (1818–1888). Jarves collected for many years, but near the end of his life distributed his collections. Some of Jarves's paintings were given to the Metropolitan Museum of Art in New York. Jarves also gave the Metropolitan Museum a collection of Venetian glass in memory of his father, Deming Jarves. Yale got 119 Italian paintings. Some of Jarves's collection was purchased by the Cleveland Museum of Art. Jarves was an accomplished writer and student of art. While not really a philanthropist, Jarves played a role in the civic appreciation of art. Art and community fascinated Libbey early on. Furthermore, art collecting was the hobby of great industrialists such as Andrew Carnegie (1835–1919), J.P. Morgan (1837–1913), John D. Rockefeller (1839–1937), and Henry Clay Frick (1849–1919). It was a way to combine interests and philanthropy. Libbey followed the career of Andrew Carnegie and his interests closely.

The real turn in Libbey's interest came with his move to Toledo. His first guide, George Pomeroy, was an art collector, but his wife Florence Scott Libbey was his greatest influence. While not initially a collector, Florence was well versed in the arts.

She and her family had been art supporters in Toledo. She had been educated in Europe and New York with an understanding of the arts. The Scotts had been visionaries, seeing a link between art and industrial crafts, and had founded the Toledo University of Arts and Trades. In addition, the Scott family had been behind Toledo's first library and park system. Florence had inherited this passion for the arts and art education, and was actively supporting a number of local artists and groups prior to 1900. Her vision was the ideal match for Libbey's evolving concept of art and the community.

The Toledo Art Museum arose from a confluence of several organizations and men. Toledo's oldest art organization was the Toledo Art Association, founded in 1883. The association offered art lessons to amateur painters, and at times, brought artists from other cities to exhibit. Many members also belonged to the Tile Club of Toledo, a group of local male artists to which Libbey belonged. The struggling artists met once a week for dinner beginning in 1895. An offshoot of the Tile Club in New York City, it was a diverse group of artists and collectors. Within a few years, the club attracted others beside artists such as architects, rug collectors, photographers, and ceramic collectors. They would cook dinners, talk, drink, and paint long into the night. Florence disapproved of these nights of drinking with the boys. Often the gang would drop Libbey off and ring the doorbell, and, fearing his wife, they ran. Still, the group was an important social outlet for Libbey.

Architect as David Stine, who designed Libbey's World Fair house, brought Libbey into the group, among wealthy artists and collectors such as Charles Ashley and Carl Spitzer, and some true artists such as Edmund Osthaus and Almon Whiting, who had gained national recognition. Some members were community developers such as Robinson Locke of *The Toledo Blade* and his wealthy friend Barton Smith. In the Tile Club Libbey met Toledo's "man for all seasons," George W. Stevens. Stevens was best known as a newspaperman, but his interests were far-reaching. He was an amateur poet, painter, Egyptologist, astronomer, storyteller, a collector of rare books, and an entertainer. Men like Stine and Stevens fostered the idea of a public museum, a project that had been in Libbey's mind for years.

Another important art stream in Toledo was the Toledo Art Loan Association, the brainchild of Robinson Locke of *The Toledo Blade*. Robinson Locke was an art collector and member of the Tile Club. The Art Loan Association Exhibit of 1898 was a charity art function for the Toledo newsboys' and bootblacks' union. Robinson Locke, who was the organization's president, put this large exhibit and sale together. Major patrons and exhibitors were Arthur Secor, W.A. Gosline, A.L. Spitzer, Clarence Brown, F.B. Shoemaker, and Carl Spitzer. Interestingly, Libbey was a patron, but had no art to loan directly. Solon O. Richardson was an exhibitor and offered many old glass pieces. Spitzer's collection offered an impressive base. Secor, Spitzer, Shoemaker, and Gosline all had extensive home collections. Robinson Locke would become one of the seven co-founders of the Toledo Museum.

Some attribute the creation of the Toledo Museum of Art to a trip the Libbeys took to Chicago. Florence Libbey's old friend Rosa Cristine Lang was a teacher at the Armour Institute there. Lang and the Libbeys visited back and forth over the years, but in April of 1900, the Libbeys visited Lang for a long weekend. Edward and Florence stayed a little longer to tour the Armour Institute and meet the institute's president, Frank Wakey Gunsaulus (1856–1921). Frank Gunsaulus was a dynamic and motivational speaker and a prominent preacher. He had become famous as a touring lecturer for the traveling Chautauqua cultural fairs. He was an art collector and a passionate supporter of the arts and education. He had impressed Chicago magnate Philip D. Armour, who gave him the money to build a "school to teach everything to everybody," which became known as the Armour Institute of Technology (today's Illinois Institute of Technology). Gunsaulus was also a founding member of the Art Institute of Chicago and the Field Museum of Chicago. He often promoted the idea that individuals should make donations to the arts. Gunsaulus had himself donated art to many institutions, such as the Art Institute and the University of Chicago. Gunsaulus had much in common with Libbey, since he had been a key member of the Chicago Fair Committee of 1893. Gunsaulus had expanded the Art Institute of Chicago for the fair, and that's when Libbey visited the institute for the first time. Collections under Gunsaulus were very broad in nature, including manuscripts, rare books, pottery, ornamental weaponry, and paintings.

Clearly, Gunsaulus's views and approaches influenced Libbey. In a letter to Rosa Lang after meeting Gunsaulus, Libbey noted:

> I little thought, when we were influenced by you to remain in Chicago a few days longer that such a few days could contain so many new pleasures for us. Through you we met the dear Dr. and caught his charming influence and also quite an amount of new thoughts, in a channel altogether new, but surprisingly mutual to us. I was in a position just then, being somewhat free from business thoughts, to grasp upon any resource much needed and to find that Florence, as well as myself, "caught on" so mutually, has brought to us more pleasure than we have ever experienced. I began to think that nothing but business could satisfy me and while I have been experimenting considerably of late trying to find something to tie to, which might, in part, compensate me for the partial loss of a business life.[6]

An art museum and community center for the arts seemed to integrate many of Libbey's interests.

Libbey had discussed his visit with Gunsaulus with the Tile Club, which generated renewed enthusiasm for an art academy. On the evening of April 10, 1901, a museum organizing committee met, consisting of Edward Libbey, Barton Smith, Edmund Osthaus, David Stine, Charles Ashley, Robinson Locke, and Almon Whiting. These men are often referred to as the seven co-founders. The main link was the Tile Club, but it was a diverse group. Edmund Osthaus and Almon Whiting were artists. Robinson Locke was a jack-of-all-arts including the theater. He collected early works of Shakespeare as well as art. Barton Smith was a friend of Locke and a fellow mason. Smith was a lawyer who had been a partner of Fredrick Geddes. Stine was an architect who had often worked for Libbey. Spitzer, Secor, and Shoemaker of the Tile Club also played important roles in the founding of the museum, and they would become major art contributors.

Formal incorporation took place on April 18, 1901. Almon Whiting, an established artist and Tile Club member, was named director. A board of fifteen trustees was established in June of 1901, with Libbey as president, and George Stevens, Charles Ashley, and Whiting as officers. There were 120 members, each paying dues of $10 per year. The stated mission of the museum was to be a gallery of art and an academy for advancing knowledge. The idea was very much in the mold of a Gunsaulus project. The first undertaking was a simple exhibition of art held through December of 1901 in rented rooms at the Gardner Building. Libbey brought the idea to the Toledo Club for support. He was able to line up some wealthy patrons in Carl Spitzer and Jay Secor, both of whom had an interest in art and were Toledo Club members. Spitzer, a Tile Club member, painter, friend and neighbor of Libbey, had a large personal art collection. His West End mansion, known as Innisfail, was a treasury of art. Innisfail had a large, two-story art gallery. Libbey had spent many hours with the Spitzer collection and had started his own collection in 1895. Locke was another driving force. Locke was the first president of the Toledo Symphony Orchestra League in 1897 and was committed to the cultural improvement of Toledo.

Andrew Carnegie had used an annual art exhibition in Pittsburgh to build interest in an art museum. The Chicago Art Institute had done the same thing. Libbey was an annual visitor to the Carnegie museum and the sales exhibition, making his first study of a functional art museum. Whiting had been working with the annual art exhibition of the Society of Western Artists for six years. Gunsaulus suggested that Libbey use Chicago art dealer Henry Reinhardt (1859–1921) to line up painters and paintings. Reinhardt was an important dealer with galleries in Chicago, New York, and Paris, and was instrumental in acquiring art for the Chicago Art Institute. He had also helped build the collection of the Detroit Museum of Art (1885). Reinhardt became an advisor to Libbey in the development of the museum. At the Carnegie Art Exhibition, Libbey met Henry Clay Frick, one of America's great collectors, who ran the exhibition.

The initial success of the Toledo exhibition led Libbey to purchase the house of P.T. Brown on Madison and 13th streets as a home for the museum. Whiting, the first director, was an artist, photographer, and instructor, although his main strength was in instruction, not administration or promotion. The project lost momentum and faltered. After two years, the museum was described as "one desk, six office chairs, one painting of sheep crossing a meadow, and one mummified cat."[7] The museum's first painting was "Sheep in Pasture" by Willem Steelink. Total assets were listed as

$285. The museum had evolved into a sort of clubhouse for the old Tile Club. One exhibition drew only seven visitors. Whiting had failed to strike a chord with the residents of Toledo, and he resigned. The museum was viewed as an exclusive club and alien to a manufacturing town. Whiting was a great collector and a fair artist, but he lacked what the museum needed most — salesmanship. Libbey realized he needed a motivator, promoter, and an administrator. He needed a Mike Owens for the museum.

George Stevens, a Tile Club member, offered sales ability. Libbey found support for appointing Stevens (1866–1926) as director from lawyer friend Charles Ashley. In Stevens, Libbey would find his Mike Owens for his struggling museum. Stevens was an eccentric, but passionate about the project, and would become Toledo's Gunsaulus. As a "builder" of the city art museum, Stevens was an Owen-like creative genius. Once again, Libbey showed his knack of finding the right man for the job, or in this case, a man and a woman, George and Nina Stevens. Nina Stevens would become assistant director. Like Owens, Stevens was an innovator, and he would breathe new life into the museum. He had studied art, but at the time was more of an entertainer. He managed a burlesque theater in Toledo and was popular at private parties for his mentalist act and storytelling. He had a mixed background that included newspaper reporting and advertising. In retrospect, the success of a museum in an industrial town depended on the unique selling abilities of the Stevenses. George Stevens was a very personable character, using nicknames such as Mr. E.D. for Libbey.

Stevens started the job immediately with an exhibition of Japanese and Dutch prints. He and Nina headed for the library, since they knew nothing about the subject. They put burlap on the walls to highlight the Japanese prints. Mrs. Libbey's uncle, Frank Scott, donated a bust of Julius Caesar, and then went to local schools promoting an opportunity to see the conqueror of all of Gaul. The Stevenses lined up some students to serve coffee and sandwiches. The success of this first effort was a real tribute to the salesmanship of George Stevens. Libbey held a banquet at the Toledo Club to reward Stevens; and at the end of the first season, Libbey took him to the exclusive Triton Fish and Game Club in Canada. (Stevens loved to trout fish.) It was a trip he remembered, and he often recalled Libbey carrying his equipment and a case of champagne many miles from Quebec. Using Indian guides, they fished and camped for days. Years later, Stevens would often be the weekend guest of Clarence Brown, Carl Spitzer, Jay Secor, and Bill Gosline at Toledo's exclusive Castalia Fishing Club.

The couple was constantly promoting the museum. They often called through the telephone directory alphabetically. Stevens called teachers, visited schools and churches, and solicited museum visitors and patrons. He had a "party" personality which made him a popular guest among the wealthy of Toledo. Stevens published the following manifesto as to pose the question, "Why an art museum?" and publish the following:

- Most men mistake being alive for living
- The best of what we hear we fail to understand
- We labor that we may gorge ourselves — and sleep with the kitchen cat and kennel dog
- A community is as rich as its understanding of the use of riches
- We grieve for the few in asylums for the mad but heed not the wanton waste of sound minds
- A great manufacturing centre is a prison house unless it supplies something for the leisure hours
- The busiest city on earth is fast asleep unless it is doing something towards the higher education of its people
- Work should be the means to leisure in which to enjoy the sublime creations of science, literature, music, and art
- No city is great unless it rests the eye, feeds the intellect, and leads people out of the bondage of the commonplace
- Hospitals do much to make men well — museums of art do more; they make WELL MEN BETTER

This powerful manifesto explained why wealthy industrialists should consider an art museum as philanthropy on a par with hospitals and social services. Stevens's first goal was to get financial support from Toledo's wealthy, but some critics thought giving to address the social needs of the city was a higher calling. Certainly Stevens summarized Edward Libbey's view of charity to the community, and the attitude of the upper class members of the Tile Club.

One of the museum's first original paintings came from Tile Club member and painter Edmund Osthaus (1858–1928). Osthaus had gained a national reputation in sporting-theme art, and the Toledo Club had purchased some of Osthaus's paintings. Osthaus was a member of the museum's executive committee, but the museum would have to move beyond local painters to be successful. On his 1901 trip to Europe, Libbey started to collect art with the museum in mind. He attended a dinner in The Hague on August 28 for artist Josef Israels and acquired his first museum-quality painting, the artist's "Daily Bread," valued at $11,750, a large sum for the time. On the same trip he purchased an Albert Neuhuys valued at $6,500. Dr. Gunsaulus seems to have inspired Libbey's interest in Dutch painters. Gunsaulus had an extensive collection of Dutch masters, and Libbey was routinely in communication with Gunsaulus over the direction of the museum's collection.

Libbey participated in the 1904 loan exhibition at the Chicago Art Institute. It was a major event and brought him into the company of other great collectors such as Henry Clay Frick of Pittsburgh. Libbey loaned several Dutch paintings, including "Merry Day in Holland" by Peter Pieters and "The Coming Storm" by Jan Hendrick

Weissenbrock. Loan exhibitions were common at the time, and Libbey would bring several to the Toledo Museum.

While Libbey thought highly of Andrew Carnegie, he did not see his Pittsburgh Art Institute as a model for Toledo's museum. First, Carnegie was not a collector of art, and Carnegie saw art simply as part of culture. Carnegie had combined his art museum with a major library and natural history museum, to bring culture to the industrial city of his making. Carnegie did not envision a community-based art museum, but he did focus on new art, giving Pittsburgh America's first museum of modern art. Carnegie's personal involvement was minimal. He turned art decisions over to collector associates such Henry Clay Frick. The institute was free to the public, but lacked programs to link it to the community. The real draw of the Carnegie Institute was its dinosaurs and Egyptian artifacts. With his libraries, Carnegie required the community to buy the books, while he built beautiful buildings to house them. Carnegie's art institute was like most art museums — Carnegie supplied the art. The paintings were philanthropic gifts. Libbey was more interested in the old Dutch masters and the emerging impressionists. Most true collectors of the period had art galleries in their homes. For his part, Libbey had a more narrow vision of art and its role in the community, but a much broader vision of what constituted an art museum. Libbey's vision evolved with the museum.

In the first couple of years, Libbey and Stevens worked with art dealer Henry Reinhardt to establish an annual art exhibition. Reinhardt had been coming to Toledo to sell art for years. Carl Spitzer, Jay Secor, and Edward Libbey had often purchased paintings from him for their homes. Reinhardt had used the Secor Hotel and was more than happy to make his annual showings at the house museum. Reinhardt would finance the exhibition and purchase refreshments from the Toledo Club. In many ways, Reinhardt was evolving as well, expanding from his gallery in the Midwest. Stevens and Reinhardt became good friends, and Reinhardt visited many times to help with the setup of the museum.

Libbey had the house remodeled to serve as both a museum and a house for the Stevenses. The couple served as housekeepers and maintenance people as well as directors and curators. George Stevens used his wood lathe in the garage to make pedestals for art pieces. He shoveled snow and cut grass besides maintaining the exhibits. The museum closed for an hour at noon, so George and Nina could catch lunch. The house even included an observatory, as Stevens was an amateur astronomer. He loved pets and had dogs, cats, a monkey, ducks, finches, a lamb, and rabbits. Carl Spitzer even tried to buy a penguin as a gift for Stevens. Later in his career, Stevens purchased a farm in Waterville for his animals. He was quickly assimilated into West End society, even though he lacked the financial means of his neighbors.

Stevens, like Libbey, was disappointed at the low number of visitors. In some months, fewer than ten people visited the museum, whose image was of an art club for the wealthy. Edward Libbey, Florence Libbey, and the Stevenses were convinced that the museum had to be more democratic. They wanted to take art to the people. Libbey was well aware that the community had little interest in art, because its members had no knowledge of it. The Libbeys wished especially to interest Toledo's young people, feeling that was the place to start with cultural education, an important part of their plan. Libbey's vision of community was clearly revolutionary. George Stevens supplied the enthusiasm to put Libbey's vision into practice. He lined up museum tours and free art classes for local school children.

It was and is no simple job to sell an art museum to an industrial city, but George and Nina were up to the task. Stevens took to the streets in search of patrons. He and his wife spent evenings calling through the telephone book. They brought teachers to the museum. Stevens invited classes to the museum on field trips, and when Libbey was in town, he spoke to the visiting schoolchildren. The Stevenses designed a number of clubs to meet the broad interests of the community. As a photographer, George Stevens established a camera club, which attracted many new members. Nina Stevens held classes on art history and gave free weekly lectures, averaging 75 people in the first month. Speakers from other cities and from universities gave talks on art and archaeology. Stevens joined the downtown Toledo Club, where he lunched with Libbey and Toledo businessmen. No one loved the atmosphere of the Toledo Club more than Stevens. He arranged for minor paintings to be donated to the Toledo Club to help promote a love of art in the city. With their array of skills, the Stevenses offered basic art classes in many materials. Wealthy community residents came to the aid of the museum as well. The Tile Club had a number of art lovers such as J. Secor, who would remain a donor and supporter of the museum throughout his life. Stevens proved to be an outstanding fundraiser at all levels of society.

Stevens used his newspaper ties and those of Tile Club member and newspaperman Robinson Locke. Stevens was a great showman, who could make a show out of a single painting. That year a New Yorker loaned him the *Russian Wedding Feast* by Makovski. It had gained some notoriety at the Chicago World's Fair and was sold commonly as a reproduction. Stevens got publicity on the painting and how he was guarding it 24 hours a day. Papers of surrounding towns wrote about it and he sent free tickets to reporters. He gave lectures on the *Russian Wedding* to educate and develop more interest. Admission was 10 cents for adults and a catalogue was 25 cents. Volunteers collecting the admission fee included board members such as Isaac Knisely, president of the Northern National Bank. To implement Libbey's vision of workers visiting the museum, Stevens set aside one night a month for factory girls to bring their escorts to the museum for coffee and doughnuts. Libbey was clearly impressed with the Stevenses' passion and effort, as well as the initial results. Stevens often booked low-cost exhibitions that were unusual and could attract press attention. One of these was the 1902 Japanese Print Exhibition, whose artwork was borrowed from one Japanese-American living in Cincinnati. Stevens and Locke always sent out press releases to major cities and often got favorable stories.

Libbey changed his mind about an admission charge in the first years of the museum. Children had always been admitted free, but the museum went to entirely free admission and sold limited memberships to patrons.

The viability of a museum in an industrial city was still a question mark. The initial novelty of the community museum wore off, and Libbey wanted a living museum of art education, a mission to which he and Stevens returned. Libbey noticed in the first year of the new museum that it attracted few students and teachers. He wrote in his first annual report: "I would suggest that hereafter we allow teachers and pupils free admission to all exhibitions upon certain days, for it is my opinion that the object of our institution, the education in and cultivation of art, can find no better field than in our public schools."[8] A year later, Libbey decided to make the museum free to the public, which was the same approach used by Andrew Carnegie in Pittsburgh. Libbey, however, did promote memberships for financial support. Stevens was able to quickly build a patron list at $10 per member.

Stevens expanded the mission of art education. He offered free lectures, not only at the museum, but at schools and community organizations. He advertised at churches, stores, factories, and community meetings. Free classes in drawing and art history were offered to increase interest. Nina Stevens found success in attracting housewives for art classes. Stevens visited Toledo schools and community gatherings to tell stories and entertain. Libbey refrained from such outreach but seemed to admire it for its results. Without Stevens, the museum project would have failed. Libbey made an optimistic appraisal in the museum's first annual report in 1902:

> We are today one of the leading nations of the world, and in all branches of industry, finance, and commercialism we are looked upon with wonder. The centers of art in Europe have justly been her pride, and our country is just beginning to appreciate the importance of art education. Nearly all cities of any size in this country have their museums and galleries which are fast becoming a matter of necessity. We owe it to ourselves, to the school children of Toledo, and to the future generations to see that our good work shall continue, that we lay a foundation so solid and complete that the future citizens of Toledo will look back upon this, our pioneer work, with praise and appreciation.[9]

Libbey made that mission part of the soul of the museum, and one can still feel it today. He saw a future for the museum that other philanthropists never dreamed of.

Libbey was as enthusiastic as Stevens about bringing schoolchildren to the museum. When a group wrote him a letter about art, Libbey responded to their teacher: "Permit the children to spend an hour some afternoon at the Museum, at which time I shall be pleased to meet them, personally."[10] This was about more than just art. Libbey truly enjoyed children. Often it was Libbey, more than his wife, who enjoyed the company of friends' children. He liked playing with their neighbor Otto Sasse's daughters, Hannah and Mia. While the art museum was also Florence's dream, it was Libbey who dreamed of educating children. The couple would often interact with children on their Sunday visits to the museum. In 1903, a children's art class quickly reached an enrollment of 125, and 30 children had to be turned away.[11] In 1903, over 15,000 children visited the small museum, which often had lines outside. Early on, the clubs, training, and education attracted people to the museum. Stevens expanded the mission to adult education by starting a women's art club and adult education classes.

Libbey now knew his mission in life. The vision was rooted deep in his and the Scott family's belief that art was the touchstone of civilization and society. Florence would bring her understanding of art and creative themes to the museum, and her vision that art education had to come first. Libbey would go beyond the achievements of Andrew Carnegie and Henry Clay Frick in his outreach to educate the community. Libbey hoped to bring art in a practical way to industrial workers. Toledo's art museum would go beyond an audience of the wealthy. The mere existence of a museum would have little impact on the community. Art education was as important as the paintings in the museum, and in particular, children had to be educated to create a new generation of art enthusiasts. Libbey had a very European view of society, but realized that America had a middle class. In his business, Libbey had modified the tastes of the middle class to emulate the tastes of the rich. He had sold cut-class for middle-class tables. Now he hoped to achieve the same thing with the museum. Later, he would promote glassmaking as a link between industrial crafts and art. The museum would become an incubator for the fusion of industrial design and art.

One problem was that the museum had at this point very little to show, but the St. Louis Exposition of 1904 would change that. As president of Libbey Glass, Libbey had been involved in preparations for St. Louis. Libbey Glass was preparing the world's largest punch bowl — a huge, 24-inch, 143-pound bowl with 23 cups, which attracted national acclaim and headlines. The estimated value in 1904 was $2,700. The bowl was crafted by John Rufus Denman and took the grand prize gold medal. It was featured on the cover of *Scientific American*. The consumer could buy a cut-glass, 14-inch punch bowl set for between $70 and $300, which was a considerable sum in a day when $400 was the yearly average income. Libbey combined art and business in St. Louis. These great world fairs were a tribute to man's ability to manufacture and create art. In many ways, the fair reflected the mission of the Toledo Museum. Libbey used the fair to sell his new bottle machines and view the many art exhibits. He gave away spun glass souvenirs, Chinese baby dolls, he visited with Adolphus Busch of St. Louis, and Busch became a stockholder in the European company formed to sell Owens's machines. As significant at the St. Louis fair was the purchase of the painting *De Profundis*, by Gustave Mosler. The painting won a bronze medal at the fair and Jay Secor bought it as a gift for the empty museum. Secor paid $1,500 for the painting, which a few years later was appraised at $10,000 (Mosler's death in 1906 caused the increase). It still can be seen in the museum. In addition, Stevens purchased a whole collection of Russian pictures at the fair for $2,000. While most considered these paintings very low museum grade, they were used to cover the empty walls. Stevens, of course, made much in his advertising about their showing at the St. Louis World Fair.

Chapter 12

The Glass Capitalist

In 1904, Libbey realized that the Owens Bottle Machine Company was a very valuable property. Most companies with an invention quickly put it to work reducing cost and gaining market share. Michael Owens wanted to do just that and argued for building a bottle manufacturing plant. Libbey had learned from his experience with the semi-automatic processes that with little capital investment, much money could be made by selling licenses and collecting royalties. He also had to consider the tough economic times, the tightening of credit and bank loans known as "the rich man's panic." The Owens bottle machine represented a major capital expenditure. Selling machines would be hard because of credit constraints, yet glass manufacturers needed cost reductions so the market was ripe. Libbey believed that by leasing machines, he could sell the labor cost savings but manufactures would not have a large capital expenditure. He also had to consider the political climate. Protectionist President William McKinley was assassinated in 1901, and Theodore Roosevelt, the trustbuster, was in the White House. Controlling a manufacturing segment would quickly bring government scrutiny. Libbey crafted the perfect strategy using license agreements and royalties. Working with board member and banker Clarence Brown, Libbey even allowed credit-restricted companies to pay with stock, which allowed Libbey to take ownership positions in many glass companies. Such arrangements would allow Libbey to control bottle prices without his companies being considered a true trust. From rent payments, royalties, and stock dividends, Libbey realized far more income than he would have by selling machines. The real genius of the Owens Bottle Machine may well have been in its brilliant marketing and licensing strategy.

The Owens bottle machine brought a revolution in glassmaking. Libbey realized the nature and impact of this revolution, but so did the glass unions. Early on, the Glass Blowers Union had banned participation of its members in any Owens development work. For the union, the automatic bottle machine not only meant the loss of jobs but a change in the fundamental structure of the workplace. They feared it would do what the Bessemer process had done in the steel industry, that is, virtually eliminate skilled craftsmen. The change in the steel industry in the 1870s downgraded the union to a trade union representing unskilled workers. The glass union hoped to preserve the semi-automatic processes, to save jobs where it could. The semiautomatic process used three men (two skilled), versus one unskilled machine operator for the Owens machine.

The future of the bottle companies didn't look much better than that of the union. If Libbey chose to enter the bottle industry, he could quickly dominate it with his cost advantage. If Libbey decided to be a machine tool manufacturer, larger companies with capital would purchase the machines and drive others out. Libbey's strategy was somewhat of a surprise but was as revolutionary as the machine itself, and perfectly crafted for the times and environment. To maintain high royalties, Libbey wanted to keep bottle prices stable, which also helped skilled laborers and competing companies.

The environment included a public outcry against industrial trusts. Trusts, better known as monopolies, boomed in the 1890s and early 1900s. Scandals and high living were constant headlines in American cities. The formation of America's first billion-dollar company, United States Steel, by J.P. Morgan had caused new fears of monopolies. President Teddy Roosevelt had promised a trust-busting administration, and was particularly focused on vertical integration, controlling the supply chain of resources for an industry. The formation of large companies by great financiers such as J.P. Morgan created a shortage of credit, making it difficult for companies to make capital investments. Libbey's strategies, however, would prove the equal of those of J.P. Morgan. In reality, Libbey controlled the bottle market more heavily than Carnegie steel or Edison electrical distribution. Libbey's strategy was revolutionary, but he had internal opposition.

In 1903, Libbey brought the board together to discuss how to proceed with the new invention. For his part, Michael Owens wanted to enter the glassmaking business and produce bottles, but he did not attend board meetings. Libbey had been looking at strategies since 1901. Selling patent rights of previous inventions had been profitable, but far less so than it could have been. Selling the rights to the bulb-making and chimney-making processes had cost Libbey in lost opportunities. For example, Libbey sold chimney rights for $100,000 and a stock position; but within a year, a new company was making far more than anticipated. Libbey made much less than he would have by direct ownership. Selling inventions to competitors had hurt Libbey Glass profits. In addition, the European company set up to sell rights to the bulb and tumbler processes had failed.

Libbey now proposed that he license the machine with royalties on production. Other inventors either used the machine to manufacture inventions themselves to gain a cost advantage, or went into the business of selling their inventions. Of course, a third, less attractive option for poor inventors was to sell the patent rights. Libbey's plan to lease was a perfect strategy for the time and business. The leasing agreements allowed a low capital investment by the lessee, and assured Libbey a long and high return. He could squeeze out future profits in an expanding market. Furthermore, he could control prices; in fact, Libbey was forced to maintain high prices. He was in essence offering the machine as a super laborer, which was paid a wage. He shared the labor cost savings of the machine by setting royalties at half the savings. The Owens Bottle Machine Company handled the licenses, and a new firm, Kent Machine Company, built the actual machines. Kent was a foundry and machine shop.

As usual, Libbey distributed his stock ownership to assure his personal control of boards and motivate directors. Kent Machine was to supply all the Owens machines to be installed, which assured financial success. Kent Machine had moved to Toledo in 1902 because of its increased business with Toledo Glass. Libbey chose to buy control of the Kent Machine Company instead of going directly into the manufacturing business. Kent Machine's major shareholders were Edward Libbey, M.J. Owens, W.S. Walbridge, and Fredrick Geddes. Kent Machine was its own entity, as Owens Bottle did not initially have the cash to buy Kent outright. Kent Machine's capital came primarily from its majority stockholders. In addition to building the machines, Kent was responsible for installation. W.S. Walbridge was the operating manager, while Owens was responsible for machine installation. Walbridge was clearly being rewarded for his skillful work in managing Toledo Glass's development project and keeping Bock and Owens in line.

The Owens machine needed specialized molds to create customer-specific bottles. Libbey, Owens, Walbridge, and a local manufacturer, Frank Woodruff, formed a foundry and machining company known as Woodruff Glass Mould. Kent Machine and Woodruff Glass Mould would make Walbridge a wealthy member of the "Toledo group." However, both Kent and Woodruff would struggle to meet the demand for the Owens machines, as the cost of the developmental work had strained all of Libbey's companies.

The first license of the machine went to Baldwin-Travis (Thatcher Manufacturing) in September of 1904. Thatcher was licensed to make milk bottles at a cost of $250,000. The company paid $150,000 cash, $50,000 in preferred stock, and $50,000 in common stock. Royalties were based on a per bottle calculation of labor savings from the machine. One-half of the savings would go to the Owens Bottle Machine Company. Libbey selected companies exclusively for beer bottles, ale bottles, wine bottles, soda pop bottles, brandy bottles, etc. The royalties in 1904 were 50 cents a gross for Thatcher. The Thatcher machine required modifications for its thick-walled, square bottle. Libbey sent Mike Owens to help in the machine start-up and installation. Owens quickly became a thorn in the side to a broad segment of the glass industry. Walbridge was forced to write the following to the president of Thatcher Manufacturing:

> Owens is brusque in many ways, but as a factory manager he has few equals and his advice is sought by a great many and as an example he is consulting superintendent of Libbey Glass Company, employing thousands of hands, and through his advice their factory is in most excellent shape. He may suggest radical changes but I should heed all he says and I believe can get you out of a hole.[1]

Libbey again stayed out of these personality and people issues, focusing on more strategic issues. Walbridge often acted as a mediator between customers and Owens. Initially, Walbridge struggled with the role; but over the years, he grew to respect Owens. Walbridge, as a brother-in-law and close friend to Libbey, never felt he was competing with Owens for Libbey's favor, and that allowed for a better relationship with Owens.

Licensing was all-important to Libbey, who sold them himself. He and Owens believed the machine could make a major contribution in longneck beer bottles. Iron City Brewing in Pittsburgh was one of the first companies to come to Toledo to see the machine. In 1903, Libbey was in negotiations with a group of Pittsburgh beer bottle manufacturers, but was getting nowhere. The manufacturers negotiated as a group, fearing that the machine would give any one of them an unfair advantage. Libbey secretly approached one of them, Edward Everett, after negotiations broke down. They exchanged letters using fictitious names. Libbey and Edward Everett joined together to take over three bottle plants in Ohio (Newark, Massillon, and Wooster). The new company became Ohio Bottle Company, with Libbey having the only Toledo group shares. Ohio Bottle Company was then given an exclusive license to make beer, porter, ale, and soda bottles. This company would be the major supplier of the Pittsburgh Brewing Company (Iron City Beer). Owens Bottle Machine Company also formed the Northwestern Bottle Company, which was a demonstration plant to manufacture bottles. It was a concession to Owens, who wanted to be active in bottle making. Northwestern employed Tom Owens, his brother, as a supervisor. Northwestern Bottle Company was given the exclusive rights for wine, brandy, and specialty bottles. Owens Bottle Machine Company sold exclusive rights to H.J. Heinz for ketchup bottles, Illinois Glass and Bolt for whiskey bottles, Hazel-Atlas for general bottles, and Ball Brothers for fruit jars. Owens Bottle Machine Company gained stock positions in all these companies. By 1905, Libbey controlled a complex web of glass companies, but avoided the appearance of a monopoly.

Libbey became America's most adept monopolist, controlling the bottle market while the government seemed unable to understand his control. With Kent Machine, Libbey had started a process of vertical integration, controlling the supply chain, but his horizontal integration of markets was overwhelming. In 1905, Libbey had control of the cut glass market and the bottle and container markets, where he was able to control pricing and production. He realized that the automatic bottle machine had the potential to cause havoc in the marketplace. Libbey was careful to slowly change the market while maintaining price levels, which would also allow for an orderly transition of glass blowers and gathers to machine operators. The union was rumored to want to buy exclusive rights to the machine for $300,000 to destroy it. Many times, security was required to protect machines from bands of neo-Luddites. Unions often prohibited workers from cooperating with companies installing machines. With a purposely-slow conversion to automation, Libbey won the support of the glass unions and the industry as a whole. His approach also won tacit approval from government watchdog groups. Still, Libbey's tactics were every bit as monopolistic as those of J.P. Morgan. Libbey, in fact, had the power to decide segment winners, because he sold exclusive rights to beer bottles, vegetable jars, milk bottles and the like.

Libbey moved quickly on expanding licensing to Europe, with trips in 1904 and 1905. Owens European Bottle Machine sold the licenses. Beer king Adolphus Busch, who had first seen the machine in 1903, became a stockholder in European Bottle Machine, whose board members were the same as those of Toledo Glass, except for Busch and Julius Prince of Germany's Appollinaris Company. Owens European had exclusive rights for Europe, Central America, South America, Africa, Cuba, and Australasia. Against some resistance, Libbey moved forward with plans to build a model bottle plant at Manchester, England. While Owens supervised the building, the board brought in William H. Boshart to manage it. Boshart was a sales manager for Pittsburgh's Cumming Glass, but lived in Toledo in an apartment house owned by Owens. Owens suggested him to Libbey for the Manchester plant. Meanwhile Bock was supervising construction of a bottling plant for Germany's Appollinaris Company. Engineer LaFrance was working on both plants, but technically Bock was LaFrance's boss. Mechanics from Toledo Glass were sent to Europe to help with the installation of the bottle machines. Libbey was also on the scene, using the time in Europe to shop for art and visit museums.

The year 1905 was one continuous struggle for the Toledo Art Museum, but Stevens was making steady inroads with the community. He was selling sustaining memberships for $50 a year and family memberships for $10 a year, and a newsletter for 50 cents a copy. Members and school children were admitted free. That year Libbey and Stevens held a very successful Impressionist paintings exhibition in Toledo. Paul Durand-Ruel (1831–1922) arranged the exhibition after meeting Stevens in Europe.

Durand-Ruel had an impressionist gallery in New York, where Libbey and Stevens met to plan the exhibition. Durand-Ruel gave the fledging museum a painting by Bethe Morisot, one of Libbey's favorites. Durand-Ruel built the country's best impressionist collections at the Metropolitan Museum of Art and the Carnegie Art Institute. Libbey was particularly interested in the paintings of the masters of the Italian Renaissance and the early impressionists. George Stevens even purchased some paintings with a long-term payback plan. The public response was a turning point and the Libbeys decided to expand the house for future exhibitions. In 1906, Libbey purchased his Rembrandt *Portrait of a Young Man* for $60,000. That same year, he purchased an Anton Mauve painting for $32,500. These prices put him in the league of Henry Clay Frick and J.P. Morgan. He donated some lesser paintings from his home to the museum. Albion Lang donated *The Ploughers*, John Hoffman donated *The Torrent*, and A. Chesbrough donated *The Lady in the Carriage*. It was a beginning, but few in the industrial city of Toledo had any idea of who Anton Mauve was. With the donations of 1905–1906, Libbey contracted to add wings to the old house museum. Lesser paintings were also donated by Frank King, Chesbrough, Robinson Locke, and Hoffman.

George and Nina Stevens continued their tireless efforts to promote the museum. They constantly solicited area schools to partner in activities. George Stevens published his newsletter, reminding members of the deadline for dues. They took every opportunity to sponsor lectures by guests, painters, and visitors. They established Sunday concerts at the museum. These free concerts included performances by high school orchestras, company bands, the Toledo Conservatory of Music, the Eurydice Club, and Orpheus Club. Saturdays saw classes in art, music, and nature. Stevens worked on a shoestring budget, but building community relationships did pay off in attendance. He tried unusual events such as a Saturday collectors fair, where children could come to the basement of the museum and trade stamps, coins, and other articles.

The most significant event of this period was the Libbey's 1906 trip to Egypt. Egypt was becoming a must-see for wealthy Americans. Newspapers followed the archaeology of the pyramids with a passion. Men like Carnegie, Frick, and Morgan were flocking to Egypt, and funding was pouring into expeditions. Libbey brought hundreds of artifacts back to Toledo. The artifacts included two mummies and two painted inner coffins (cartonnages). In late 1906, the museum opened an Egyptian gallery, which became wildly popular with the public. Nothing did more for the popularity of the museum than the mummies. The gallery drew people who had never dreamed of entering an art museum. In 1906, the Ta-Mit mummy exhibit was advertised as "opened, disclosing the mummy which has been partly unwrapped to reveal the remarkable preservation of the body which has been in its present condition for over twenty years." A conductor on the Central Railroad told the local paper that his little boy dragged him to the museum. "Since the lad had heard there was a real mummy to be seen in the Toledo Museum of Art, the father had been given no peace until the visit was paid."[2]

Libbey's and Stevens's love of the mummies was noted years later by Nina Stevens:

> When the cases were opened George Stevens and Edward Libbey were like a couple of school boys. They searched for treasure in the bosom of Mrs. Tamat [Ta-Mit mummy], with appropriate and inappropriate remarks, but alas, the wrapping had been displaced, perhaps by the Turks some five hundred years before, and the jewels taken. They named

> the one mummy Willy, because as he was unwrapped, he fell into George's arms and was taken for a dance around the gallery, while George sang a new song, "Waltz Me Around, Again Willy." Mrs. Libbey and I were on the side lines laughing, in spite of womanly fears lest our husbands might be catching an ancient germ or two.[3] Another interesting point of these early museum years was the closeness between the Libbeys and Stevenses, who seemed to be able to put the formal Florence Libbey at ease.

Another object in the 1906 collection was a marriage contract on papyrus. It would become known to Egyptologists as the "Papyrus Libbey." A 1906 *Toledo Blade* article noted: "His collection embraces examples of the work of every Dynasty from B.C. 5,000 to B.C. 300, truly a remarkable gathering of intensely interesting objects." The crowds overwhelmed the museum, averaging 2,000 a month on a very limited schedule of hours. And the exhibit brought national attention to Toledo.

The museum attracted the money of wealthy art lovers such as C.S. Ashley, George Pomeroy, Edmund Osthaus, Robinson Locke, Arthur Secor, C.A. Secor, Frederick. B. Shoemaker, and W.A. Gosline. Florence Libbey and the Scott family were prime movers, as were Libbey executives such as Clarence Brown, F. Geddes, J.D. Robinson, and Solon Richardson. Even Mike Owens was a sustaining member, but he had no other involvement. Interestingly and ironically, several of these glass executives served on a national committee of museums seeking to remove tariffs and duties from paintings and art objects. The committee included another millionaire who profited from American tariffs, Henry Clay Frick of United States Steel. The committee argued that learning design from imported art would help America maintain a manufacturing edge, and in the future, allow a reduction on other tariffs. The committee said, "Design in our manufactured goods would soon show the effect of such art education, and America would longer be handicapped in this respect."[4]

The success of his glass companies and the progress of the museum allowed for a type of routine to set into the Libbeys' lives. They lived a peaceful life, below the radar of a national press that was following the parties and high living of New York capitalists. The Libbeys vacationed in Pasadena, California, and lived quietly in their West End neighborhood. They often summered in Europe, where Libbey purchased the majority of his art in Paris. While many of their neighbors went their separate ways in Europe, they would often meet in Paris during August for some serious collecting. The Libbeys had a New York apartment for shorter trips to the East Coast. As the years went on, they spent more time there. Florence's old family friends and neighbors dominated the Libbeys' circle of friends. Typically, they entertained neighbors such as Dr. and Mrs. Otto Sasse, Mr. and Mrs. William Gosline, Mr. and Mrs. Carl Spitzer (he was the son of A.L. Spitzer), Mr. and Mrs. Ketcham, Mr. and Mrs. Secor, Sarah Libbey (Edward's sister), Mr. and Mrs. George Stevens, and Mr. and Mrs. W.S. Walbridge. William Gosline was a major regional coal dealer, and a good social friend of Edward Libbey. Gosline's son Bill was also a member of the neighborhood group, who became very active in the administration of the museum. The Libbeys, without children, were very fond of Otto Sasse's two daughters, Hannah and Mia. The Sasse daughters called Libbey "Uncle Ed." The Libbeys loved dogs and cats and the neighborhood children enjoyed them too. Dr. Otto Sasse was the personal physician of many in the West End, including Edward Libbey, and he and his wife were the Libbeys' closest friends. Edward Libbey was very playful with the children, while Florence was a stickler for good behavior.

The Libbeys always decorated the Sasses' Christmas tree on Christmas Eve, and the Sasses would spend New Year's with the Libbeys, who often held the neighborhood New Year's Eve party. In the summer, the Libbeys would go to the Sasses' cottage at Put-In-Bay. In earlier years, they might all go to Put-In-Bay on Solon O. Richardson's yacht. Florence Libbey had several close friends in Rosa Lang, Mrs. Otto Sasse, and Nettie Poe Ketcham (1865–1950). Nettie Poe Ketcham was a distant cousin of Edgar Allen Poe. Her father was a founder and director of Toledo Savings Bank & Trust. She was considered the "belle of Toledo," and her parties were major social events. After the death of her husband, John Ketcham, in 1898, she moved to New York and maintained a summer cottage in Heimberg, Switzerland, where both of the Libbeys routinely visited her. Florence Libbey made many trips to New York to shop with Nettie Ketcham, and spent weeks at her Swiss cottage as well. Both Rosa Lang and Nettie Poe Ketcham were interested in art and art education for children. Ketcham became a supporter of the Toledo Art Museum. Later, she would become a substantial contributor to the museum's School of Design.

Living in the beautiful West End was comfortable without being ostentatious. The area was roughly a triangle formed by Detroit Avenue on the west, the Scott property (the museum) on the south, and Collingwood Boulevard to the east. The Scott family had owned most of the land originally and had promoted the care of the woods. The residents were Toledo's wealthiest, who had beautiful homes with maids, butlers, cooks, and stable attendants. The Libbeys had a personal staff of seven. Toledo's best architects, such as Edward Fallis and David Stine, designed the houses. Dinner parties were a common and popular form of entertainment in the 1890s. The woods were maintained beautifully, both by the Scotts and by the Macomber family, their good friends. Aaron Macomber was the Johnny Appleseed of Toledo, planting trees along streets as a personal beautification project. His care is reflected in the word "wood" in many of the street names. Collingwood Avenue, lined with old elms, was the equivalent of New York's Fifth Avenue or Pittsburgh's East End. The churches were imposing, such as the First Congregational Church, the First Unitarian Church, the Collingwood Avenue Presbyterian Church, Rosary Cathedral, and St. Mark's Episcopal Church. For leisure-time activities, there was the Toledo Outing Club, which had riding paths and tennis courts. Many of the neighbors had gardens and home conservatories. Aaron Chesbrough had a conservatory built to house a full-size palm tree that he had grown from a small plant.

The West Enders were quiet people but loved art, horses, automobiles, and dinner parties. Their dinner parties often rivaled those in New York or Europe. Carl Spitzer was a world-renowned gourmet and had an elaborate wine cellar in the West End. He had even been authorized by Waldorf-Astoria to purchase full vintages while he was in Europe. The wealthy residents of the West End preferred French food, and most had their own chefs. Aaron Chesbrough was known for his gold dinner plates and his third-floor ballroom. Libbey was one of the first members of the Toledo Polo Club, and loved riding horses his whole life.

The wide avenues of the West End were favorites for the new-fangled automobiles, and Sunday drives were like a parade. Spitzer had the first automobile in Toledo. His $50,000 Napier with silver trimmings was outfitted with a chauffeur in a silvered-trimmed uniform. The Spitzers had one of the first Oldsmobiles, and an Electric Victoria. Carl Spitzer was known for his 27,000-mile automobile trip across Europe in 1909. Edna Spitzer would only drive the Victoria. Spitzer kept a Rolls Royce in Europe for his trips there. Neighbors like Libbey would often join him in Europe for cross-country auto trips. George Stevens had one of the first air-cooled Franklins. Many West Enders drove Toledo's earliest gasoline car, the 1902 Pope-Toledo. Mike Owens had a racecar driver to chauffeur his Pierce Arrow and 12-cylinder Packard. Owens had many cars but would only personally drive his Detroit Electrics. He owned a Milburn Electric which he loved to drive around the West End. Electrics were the most popular automobiles in the first decade of the century. Sunday driving parties were common in the West End.

Many great art collections were assembled in the West End. When Jean Guiffrey of the Louvre in Paris visited the museum in 1914, he took the time to visit the local private art collections. He visited both Libbey's and John N. Wileys's private collections of Dutch masters; both had some of the world's best Rembrandts. John Wileys had only started his collection in 1913, after finding inspiration as his role as a museum trustee. In his first trip to Reinhardt's New York gallery, Wileys spent $1 million. In 1914, he purchased *Portrait of Lady Ramsay* by Hans Memling for $100,000, and *Portrait of Joseph Coeysmans* by Frans Hals for $150,000, putting Wileys in the league of America's greatest collector, Henry Clay Frick.[5] Other world-class paintings in the West End were the Hals in the Libbey collection and the Corots in the collection of Arthur Secor. Guiffrey also visited the collections of Pomeroy, Chesbrough, Spitzer, and Shoemaker. Carl Spitzer had a world reputation.

Very few neighborhoods outside of New York's Fifth Avenue and Pittsburgh's East End could boast such collections. Many of these great pieces of art would eventually make their way to the Toledo Museum. As trustees of the museum, Arthur Secor and Carl Spitzer often purchased art for the museum in Europe and were well known as international collectors. Owens tried to get into art, but had no feeling for it. At one point, trying to keep up with the neighbors, Owens purchased a Corot painting for $25,000. He soon decided he wanted more out of his money and returned the painting to the dealer, replacing it with a less expensive picture.

Another favorite Sunday activity in the neighborhood was to drive up the Maumee River for a picnic. The Secors, Bells, Stevenses, Goodbodys, Spitzers, Goslines and Libbeys when they were in town loved to picnic together around the Waterville area. Eventually, the Stevenses built a home on the river and stocked the river with black bass, a popular fish among the wealthy. This same group liked to weekend in Chicago for shopping and museum visiting. Mike Owens belonged to a very different social circle and preferred to go to his cottage at Clark Lake in Michigan. Many West Enders had camps and cottages in Southern Michigan to escape the summer heat of their mansions. While most of the neighborhood took summer trips to Europe, Owens preferred to vacation in Atlantic City, although he did go with neighbors in the winter to Florida to golf. Another passion of Libbey and the West Enders was trout fishing. Most belonged to a trout fishing club in Michigan known as Castalia, as well as the Skookum Club near Cadillac, Michigan. Stevens, Gosline, the Spitzers, and Brown were all passionate about fishing, and Libbey often joined them. They sometimes spent summer weekends at the Skookum Club, but this was not one of Florence Libbey's favorite activities. As the years progressed, the Stevenses, Goslines, Browns, and Spitzers made up a social circle, while the Libbeys were more drawn to California and Europe. The Libbeys often invited their old neighbors to their California home, but their roots were always deep in Toledo.

The Libbeys routinely spent winters in Pasadena, California, and took summer visits to Europe. The Spitzers and the Chesbroughs also enjoyed winter trips to Pasadena. While at their apartment in New York, Florence visited Nettie Ketcham and Libbey conducted business and collected art. Libbey loved the New York art community and the museums, but he also started to get more directly involved in the Toledo community. Libbey was a Rotarian and was pulled into many activities by his brother-in-law, W.S. Walbridge. Libbey was active in two important social service groups, the Benevolence Committee of the Chamber of Commerce and the Toledo Federation of Friends (Federation of Charities). The Toledo Federation of Friends had started in 1904, with founding members including Libbey, Arthur Secor, Carl Spitzer, and Virginia Secor Stranahan. It was part of the progressive movement that was sweeping the country. The federation maintained shelters for homeless men and a camp for underprivileged children, which Libbey had proposed. The federation also helped the unemployed and coordinated help between government and private donors. Libbey served as president of the federation from 1909 to 1912. The federation work was consistent with Libbey's vision of community.

The problem was that Libbey had little time for the routine of community organizations. He was not particularly active in the church, but was a member of the Masonic Lodge.[6] Libbey does not appear to have been extremely active in the Masons, but did a good deal of work under the lodge's president Robinson Locke during 1913–1916. Libbey also donated to the Rotary Club and Chamber of Commerce because of his brother-in-law. Clearly, however, Libbey saw the museum as his major means of investing in the community.

Chapter 13

Europe and Expansion

The year 1907 would be significant in the life of Edward D. Libbey. He would expand his European operation and marketing of the bottle machine. In the background, Libbey Glass had a steady, sometimes dramatic, increase in profits. Libbey Glass was still dominant in glass tableware, lamps, and cut glass. Consumer spending helped profits soar. Libbey Glass had undertaken vertical integration into mold making and sand, but didn't fully pursue it. The success of Toledo Glass's development work was not lost on Libbey Glass's management. While Libbey was the main stockholder in Libbey Glass, it functioned as a separate entity from his other companies such as Owens Bottle and Toledo Glass. The officers of Libbey Glass, Solon O. Richardson, J.D. Robinson, and W. Donovan, were barely on speaking terms with Mike Owens and wanted nothing to do with the Toledo Glass Company. In fact, they believed Toledo Glass had overcharged them for technology that they helped to develop. Owens, at Libbey's suggestion, was a Libbey Glass consultant, but Richardson wanted to start his own research and development company. Again Libbey became the lynchpin of several competing organizations. As time passed, Libbey had more difficulty in maintaining friendships among the two competing factions. He remained president of Libbey Glass but reduced his stock position in the company.

The booming light bulb business of Libbey Glass was attacked by foreign companies with cheap labor. In addition, domestic competition had increased, as electrical lighting became America's premier growth industry. Libbey Glass had up to 40 percent of the market, with another 40 percent share going to Corning. Corning was already working on a fully automated glassmaking machine and had the financial backing of J.P. Morgan in New York. General Electric in 1895 had split the market 40-40-20, with Phoenix Glass of Rochester, Pennsylvania, taking the 20 percent G.E. share. The agreement called for a 21 percent return above production costs, but expired in 1910. J.P. Morgan, who controlled General Electric, had pushed for this market sharing, and Libbey concurred, but Teddy Roosevelt's administration was aggressively reviewing such monopolistic behavior. Libbey Glass was using the Owens semi-automatic paste system to produce bulbs, which still was highly labor sensitive.

Richardson moved ahead with tacit approval from Libbey to form Westlake Machine Company. Libbey Glass and its stockholders financed the new company. The capitalization of the company in 1907 was a mere $10,000, but Libbey Glass stockholders would add operating funds. A small experimental shop was set up on Westlake Street in Toledo. The mission of the company was open-ended, by Richardson's design, "to design, develop, demonstrate, manufacture, buy, sell, lease, or otherwise acquire and handle automatic machines of all kinds." The initial focus was the development of a fully automatic bulb machine. Libbey realized, as before, that such a machine offered a huge potential for financial gain. Libbey was president of Toledo Glass, Libbey Glass, and Owens Bottle, holding together a fragile alliance. Owens and Bock would act as consultants, but at the time they were both tied up in Europe with bottle production. August Kadow of the Owens Bottle Company was brought in to directly manage the operation. Libbey's friend, Chas West, was named president. Kadow had ideas of his own from Owens's design, and Richardson agreed, wanting a machine independent of Owens. Richardson had always felt his ideas had been incorporated into the Owens machine with little credit to him. He planned to use the Owens suction-gathering principle with a new type of machine. For his part, Owens forced Libbey to impose an agreement covering Owens patents, and all Westlake patents would go to Owens Bottle. Richardson objected but was overruled by Libbey. Libbey had his hands full with Richardson, Owens, Kadow, and Bock.

Luckily for Richardson and Kadow, Owens and Bock were tied up in Europe developing bottle machine sales and building demonstration bottle plants. A contract with the mineral water conglomerate Rhein-Ahr had caused profits to soar. European bottled mineral water was all the rage in Europe and America. President William McKinley would always have a bottle of Apollinaris water with his meals. The contract called for the installation of six Owens machines. Rhein-Ahr would pay $7,000 plus freight and all duties from Toledo, Ohio. The license required Rhein-Ahr to pay 36 cents per gross or a quarter of a cent per bottle produced. While the benefits to Toledo Glass seemed very favorable, Rhein-Ahr gained a major marketing advantage in Europe. The Manchester demonstration plant was just as successful in attracting world attention. In 1907, German bottle makers and users met with Libbey to form a cartel to control Europe. The group known as Verband (Europaischer Verband des Flaschenfabriken Gesellschaft mit beschrankter Haftung) would purchase exclusive rights to the Owens machines for $3 million. This was a massive business deal for the time and would return huge dividends to stockholders of the companies connected with the machine. The license included all countries except the United States, Canada, Mexico, Japan, and China. Several years later, the Verband obtained the rights for Japan and China. The deal included all the equipment at the Manchester plant, which would be shipped to France. Owens European would pay Toledo Glass for all future machines, and issue royalties to machine users. Libbey agreed to offer technical assistance and upgraded technology as needed in the future. With the deal, Owens European Bottle Company had finished its work, except to find European machine companies to produce the Owens machine. Libbey made the most of his European venture by combining business with art exhibitions.

Libbey's personal wealth arising from the European deals helped spur significant museum improvements in 1908. Libbey responded to the pleas of museum director Stevens for a bigger building, believing that the museum's size was limiting his ability to have paintings donated. The Egyptian exhibits had people waiting in long lines to get into the small house. Stevens noted that over 20,000 children had taken art classes, and he was out of room. On Saturdays, the lines might be over a block long. In 1907, Libbey offered $50,000 toward a building and $35,000 worth of property if the community would raise $50,000. By May 1908, the community had done it. Three big donors gave $5,000 each, but most of the donations were small. Ten thousand schoolchildren contributed $500 in pennies. Stevens was behind these smaller contributions. In all, there were 549 donors, and 63 percent gave less than $10, reflecting the depth of the community support. Libbey increased his donations to $200,000, which included most of his personal art. These included paintings by Neuhuys, as well as Diaz, Tromp, and Jacque. Florence Libbey donated the West End property of her father (Maurice Scott) on Monroe Street at Scottwood Avenue. All of this was accomplished during the recessionary period known as the Panic of 1907.

Stevens extended the role of clubs and local painters. The Athena Society was added with a mission to "encourage women in all branches of fine and industrial arts." Athena membership was juried. A Business Girls' Club formed in 1907 dealt with art and business topics. The Tile Club extended its work with the Toledo Museum, producing a special exhibit on the antiquities of Ohio and Lucas County. Interest in Ohio Indian mounds and Native American art was high, and Libbey friends Carl Spitzer and George Ketcham were collectors. This exhibit proved very popular, showcasing primitive scrapers and arrowheads found in Lucas County. The Spitzer family donated a collection of pre–Columbian pottery which had been purchased for the Peabody Museum at Yale. Mrs. Robinson Locke offered classes on furniture and furniture design.

Libbey took a personal interest in the design of the museum, having studied most of the great museums of Europe. He made several visits to the Carnegie Art Institute in Pittsburgh to study its growth since 1895. The architectural firm of Green & Wicks of Buffalo, New York, was hired to prepare drawings for the Toledo Museum. The design was to be the Greek Ionic style of the time of Pericles. The entrance would be a portico of sixteen fluted columns of Vermont marble. There would be a fountain and small pool in the front. The building would have a sculpture court, a reference library, twelve galleries, and an auditorium for lectures, along with many rooms for club meetings and classes. This original building is the center portion of the museum today. With the building set, Libbey spent more of his time in Europe, collecting for the museum. Unfortunately, the national economy was locked in a deep recession during 1907, which kept Libbey's enthusiasm in check. The museum had adopted an October to May season. Summer months were reserved for the Stevenses' and Libbeys' art collecting trips to Europe. The couples would go their own ways, but before leaving Europe, they would meet in Paris with art dealers.

Stevens continued to solicit art for the bare walls, and acquired some significant donations such as *Effet de Lumière* by Jules Grun and *The First Step* by French painter Adolphie Jourdan. The board of trustees donated *Potier à Saint-Amand* by Nils Forsberg. The potter was an interesting selection by the trustees. It clearly demonstrated man the maker, showing a potter at his wheel in front of the kiln. The painting defined what was evolving as the museum's mission, and had been a medal winner in Paris the year before. Aaron Chesbrough donated *Laying the Foundation of the Kremlin*, a Russian painting that had been popular at the 1904 St. Louis Exposition. Chesbrough was a key trustee in the early days of the museum. He came from one of Toledo's wealthiest families and was a lumber and a real estate baron. Chesbrough traveled extensively in Europe and was a well-known art collector in many venues. His personal art collection had been a part of the Toledo Art Loan Association exhibits of the 1890s. Solon O. Richardson donated a cast by the British Museum of the Rosetta Stone, which became a very popular item for the museum. While in England for the bottle company, Libbey often shopped the salons for current art, such as a group of watercolors which he purchased for $372. Stevens continued to develop a huge art reference library for the museum as well. When he failed to get a painting donated, he often would get a reference book. The Art Reference Library was free to the public daily to promote the study of art. By the end of 1907, the museum had "twenty-eight paintings, forty sculptural casts, and 233 Egyptian exhibits." It was holding a number of exhibits of contemporary painters and agents which generated donations, a ploy pioneered by the Carnegie Art Institute in Pittsburgh.

Libbey and Stevens, however, were creating a new model for community museums which attracted worldwide acclaim. Libbey wanted a true community partnership. The larger American philanthropic museums, however, found some fault in this new approach. The American Art News criticized the Toledo Museum for its use of advertising in its newsletter: "The custom of soliciting advertising for museum and gallery organs and publications which was started by Director Kurtz of the Albright Gallery, and which it would appear the Museum News [Toledo] is following, would seem to be an undignified one."

The museum in 1907 had a notable group of prominent Toledoans who contributed $50 a month. Sustaining members included Charles Ashley, Harold Ashley, Clarence Brown, James Brailey, C. Braun, S. Carr, Dr. W. Chapman, Aaron Chesbrough, Frank Cheney, Judge Doyle, Edward Ford, William Gosline, Frederick Geddes, James

Hodge, William Hardee, Frank King, Isaac Knisely, Harry King, Robinson Locke, Edward Libbey, Martha Locke, Florence Libbey, Albion Lang, John Newton, Mike Owens, Edmund Osthaus, George Pomeroy, James Pilliod, Jefferson D. Robinson, Solon Richardson, C. Reynolds, S. Reynolds, F. Shoemaker, D.L. Stine, Barton Smith, Charles Arthur Secor, J. Secor, Noah Swayne, Mrs. Shoemaker, A. Tillinghast, T. Tracy, and W. Walding. Libbey was aware that his approach was different. Carnegie's museum was a pure gift, and the St. Louis Museum of Fine Arts was supported by a yearly community tax of $100,000. In fact, most big city museums got tax help, such as the annual $150,000 for the Metropolitan Museum of New York.

Libbey's view of the museum evolved as the museum progressed. His own ideas were combined with those of Dr. Gunsaulus and George Stevens, as both envisioned an art museum much different than any in the world. Stevens's wife would later write about his dream:

> A useful museum of art, he thought, should provide the mind and the hand, by intelligent analysis of the works of the masters of all ages, from prehistoric times to the present, with the means of developing a sense of proportion and color. He meant ordinary minds and hands not necessarily those with predilections for art. The hands that fashion the common things in life, the homes, the gardens, the clothes, the minds that conceive the designs of things in use, pots, the kettles, the bottles, on and up through furnishings, facades, fences, boulevards, parks, and city plans, were what he meant."[1]

While Nina Stevens was looking back, this represented what the Toledo Museum of Art would become, as well as Libbey's early view of what it could be.

Libbey's many projects in 1907 took their toll on Toledo Glass Company's focus on progress. The business success in Europe and the drain on the time of Libbey and his key managers had masked the tortured progress of the bottle machine in the United States. Owens bottles made up less than 5 percent of all bottles produced in 1907. Unions had restricted the use of Owens machines by whatever means possible. They allowed reduced wages for workers in plants using semi-automatic machines to save jobs. Neo-Luddite workers even destroyed machines at Ohio Bottle Company's Newark, Ohio, plant. Glass workers in Toledo refused to work at Toledo Glass and Owens Bottle Company. Owens Bottle used unskilled, non-union workers to man its machines. Libbey tried to control prices by adding controls to the licensing agreements. In general, the union found little success in stopping automation. The bottle manufacturers tried hard to train skilled gatherers as machine operators. Bottle makers and Libbey machine companies attempted to control prices, afraid that the machines would cause a major price decline in bottles and erode the profits of the bottle makers.

Libbey's plan with the bottle makers and licensees was to allow the reduced cost of the automation to go directly to the bottom line by maintaining prices. Through 1907, Libbey had achieved his goal; but in reality, Libbey had little control of the market. Most of the problem was anticipation of automation. The nation was in the recession known as the Panic of 1907, but there were positive forces at work too, such as the Pure Food Act of 1906. The Pure Food Act required uniform bottle size, favoring the Owens machines, which could run 24 hours a day for 12 months. The summer stoppage was becoming history in the bottle industry.

The Panic of 1907, while a short (less than three years) and deep recession, created havoc in the business world. Major companies such as Westinghouse Electric failed, as well as over 200 banks. The stock exchange was closed. Libbey's prudent management meant that all his companies were safe. However, Toledo was struggling, with several banks ready to go under. While most of the nation's banks were calling in loans on manufacturing, Libbey convened a meeting of local industrialists to save the Toledo banks. He was well aware of the interrelationship among his company and the museum with the economic health of the local community. Libbey made little progress till he got the support and money of friend and glassmaker Edward Ford.

The Panic of 1907 created a capital shortage for most businesses, but Libbey was always ready for creative solutions. The downturn in the overall business climate reduced the demand for bottles and the royalties going to Owens Bottle Company. Owens Bottle had been reporting monthly deficits. The problem seemed to be the high royalties, which concerned some bottlers. Libbey and Owens had a low-cost semi-automatic machine made in England by United Bottle Machine to provide an alternative. Improved semi-automatic processes required little capital, and union wage reductions made them a viable option. In 1907, there were less than twenty Owens bottle machines in operation. Only three of the more than one hundred U.S. bottle companies had exclusive licenses. The unions supported the installation of semi-automatics with reduced wages and special concessions to help save at least some jobs. With wage cuts and incentive pay, the semi-automatics could compete. They even offered a slight advantage in varying the size of containers, especially with wide-mouth bottles. The Owens machine required time-consuming mold changes.

Libbey felt design improvements on the Owens machine could help license sales. Mike Owens argued that they should move into bottle production instead. There were many stormy board meetings with Owens arguing for a bottle plant against Geddes, Brown, and Walbridge. Owens personally tried to convince Libbey, but Libbey stayed with the majority opinion. He remained focused on licenses, but he did allow Owens to increase bottle production at Northwestern Bottle. Owens understood that the real money in bottles was in producing them, and he slowly won Walbridge over. Libbey would come to the same conclusion within a few years, but he firmly believed the production rate and capabilities of the automatic machine needed to be expanded first. Competing semi-automatic processes were making gains because the union was willing to cut wages to keep the Owens bottle machine out of a plant. In the meantime, Libbey looked for new markets. He was well aware that increased ketchup bottle production had gotten the attention of H.J. Heinz, who wanted a license agreement. He needed sales as well as a machine development program to stay competitive.

Returning from Europe, Libbey focused on selling more licenses, and in 1907 he added twenty-two new licensees. But the European effort had cost him, with some licensees letting options expire. With the European issues settled, Owens and Bock once again turned to machine development at Toledo Glass. The new machines had some early problems, such as cooling distortions. In addition, Libbey and the Toledo Glass board wanted more productive machines to defend the market against competitors. The board and company manager W.S. Walbridge also started development work to expand the bottle size that could be made on the machine. In 1908, Toledo Glass released a new model known as AC which addressed the cooling problem and increased output by 20 percent. This machine was a real success and expanded license agreements to 55 by the end of 1909.

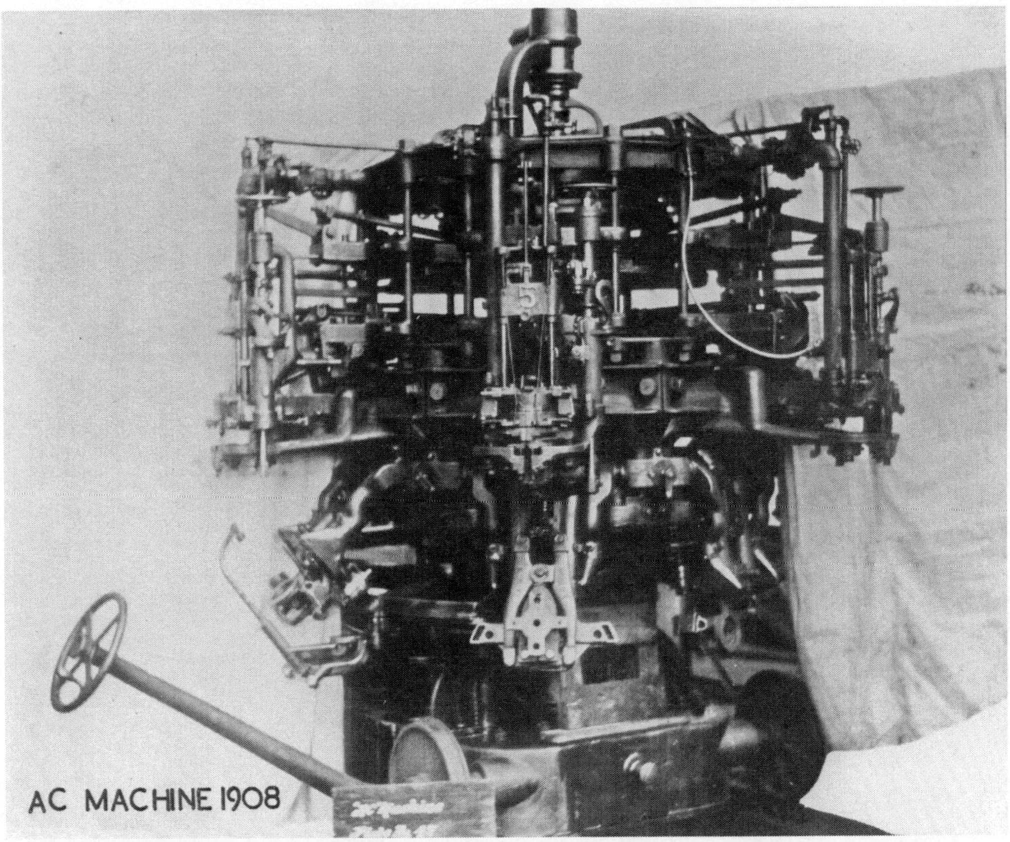

Early, developmental Owens machine. (courtesy of University of Toledo, Ward Canaday Center, Libbey Glass Records).

The year 1909 was a banner one for new licensees. Libbey added licenses with H.J. Heinz to various jars. Heinz had four Owens machines running in its Sharpsburg, Pennsylvania, plant. Ball Brothers of Muncie, Indiana, added machines to make fruit jars. Whitney Glass Works of New Jersey obtained exclusive rights for druggists' ware. Charles Boldt obtained the rights for whiskey bottles. Toledo Glass, with Owens and Bock back, turned to other innovations such as an automated pressed glass machine for making tumblers. They only made one, for Hazel-Atlas.

In late 1909, the AD model was introduced, with ten heads (molds) instead of six. In 1910, Emil Bock left the company after years of fighting with Owens. Libbey had managed to hold the pair together during the critical years, with his skill at making employees want to perform at their best. Bock was the driving force in standardization and interchangeable parts, which allowed Libbey to have machines manufactured around the world. Many on the Toledo Glass board continued to credit Bock with the invention of the machine, but Libbey finally limited such talk. For his part, Bock went on to invent the tapered roller bearing that advanced world machinery production. He later lost millions in a failed effort to build a washer/dryer combination. Libbey and Walbridge promoted Richard LaFrance to replace Bock. Libbey always had an understudy for every key employee.

Owens and LaFrance continued the line of improvements and formed a solid team. LaFrance was a good replacement for Bock and supplied the engineering expertise to improve the machines. New machines produced a larger variety of sizes and allowed for different sizes to be made on one setup. By 1912, machines could produce 50 bottles per minute, a great improvement over the 12 bottles per minute of the first machine. Later in 1912, the new eight-arm AL machine could produce one-gallon jugs. Owens and Toledo Glass also worked on continuous annealing operations to further eliminate the need for boys in the process, and started selling continuous anneals with their machines. Owens even worked with H.J. Heinz on exploring continuous bottle-filling operations. By 1912, the Owens bottle machine had dramatically reduced the labor demands in the industry as well as the need for child labor.

Libbey lowered royalties to increase market penetration and cut back exclusive rights contracts in 1910. Royalties were reduced from 40 cents to 15 cents per gross on milk bottles. Libbey moved to cash deals as well, requiring no royalties. Hazel-Atlas used $500,000 in stock and dividends to replace the royalties in the contract. Ball Brothers paid $500,000 in cash. In reducing milk bottle royalties, Libbey gained some freedom from exclusive rights, which were limiting penetration of the Owens machine. Manufacturers with exclusive rights feared a price decline with the proliferation of the Owens machine. Libbey again demonstrated his command of finance and marketing, changing policy according to conditions. In 1909, there were 51 bottle machines in operation and by 1913 there were 161 machines producing over 360 million bottles per year.

Libbey's bottle industry strategy shifted significantly after a court decision upholding the Sherman Anti-Trust Act, which meant Libbey's contracts could no longer control prices. Libbey considered moving directly into bottle-making operations, which was what Mike Owens had wanted for years. The board of Owens Bottle voted to build a new bottle-making plant at Fairmont, West Virginia, and to create Owens West Virginia Bottle. The Fairmont plant would have three continuous furnaces and six bottle machines. Mike Owens picked Fairmont to take advantage of cheap natural gas, but valley winds interfered with furnace operations, and spring floods destroyed the furnaces. Libbey and Walbridge had allowed Owens too much freedom in the decision. Libbey was slow to fully appreciate Owens's limitations. He was a plant operations man. With the museum and travel, Libbey was forced to delegate decision-making. Owen's Fairmont decision clearly lacked the methodical approach Libbey used in choosing Toledo. Libbey had another problem — his exclusive rights contracts would limit the type and number of bottles he could produce on his own machines.

One of the reasons that Libbey avoided the scrutiny of the government's antitrust lawyers was his corporate image. Glass prices had been steady and even declined with the automatic bottle machine. This price pattern satisfied the government, but what the government didn't realize was that Libbey's monopoly lowered prices for his own benefit. Libbey and his lawyers, like Brown and Geddes, moved cautiously into the bottle market. They initially controlled the market through royalties and licenses, but then they moved into bottle production at Fairmont. Libbey's companies were supplying high-paying jobs for Americans without the labor problems found in other industries such as steel. Libbey and his companies always got positive press. In 1913, Libbey received a letter from the National Child Labor Committee of New York which announced that the Owens bottle machine had done more to eliminate child labor than years of legislation. Libbey used the letter in his annual report. It was an ancillary result of the bottle machine, but it was a revolutionary change in the glass industry. Mike Owens, himself a child laborer, believed in ten-year-old children working in glass factories.

Owens voiced his opinions very strongly in a 1922 interview:

> A great deal of the trouble today is with the mothers. When I was a youngster I wanted to work.... Too many boys are being brought up by sentimental women. The first fifteen or twenty years of their lives are spent in playing.... That's all they do; at least, many of them. When they finally start to work, they are so useless and so helpless that is positively pathetic.... Work never hurt anyone! The conditions under which you work may hurt you. You will suffer if you don't get enough food and sleep for rebuilding and the rest you need. Worrying over work will wear you out. But the hard work I did as a boy never injured me."[2]

Libbey, of course, had a different view, favoring early education, although Libbey, too, had been a young apprentice.

Child labor in the late 1800s was a serious problem. Estimates suggest that 25 percent of American workers were boys under 17 years old, which was about the percentage in the bottle and pressed glass industry. The window and plate glass industry used only about 4 percent. In general, states set a minimum age of ten for working, and twelve for coal mining. By the 1880s, Ohio set the age at twelve and restricted work to ten hours per day. The state required parents to sign a certificate. State inspectors were often paid off or simply overlooked violations, but Libbey's record was outstanding. Libbey and Owens both believed that a working boy should be given the opportunity to move up in the glass crew hierarchy. Still, abuses by parents, factory owners, and state inspectors were widespread.

A gang or crew in handmade bottle making required up to seven boys, known as blower-dogs. The stick-boy or middle boy reheated the piece in the glory hole for the blower to continue working on it. The mold boy fixed, held, and washed the mold for the bottle blower. The snap-boy helped in breaking the bottle off the pontil by applying a water-cooled rod to a contact point, allowing the bottle to be snapped off. Two boys, carry-in and carry-out boys, were used to take bottles to and from the annealing furnaces. The lip-boy would grind the lip of the bottle to customer specifications. He had the most dangerous of the jobs because he used an abrasive wheel. In more complicated pieces, a bit-boy might be used to help apply things like handles. These boys worked two pots of glass each day. That required a ten-hour day in two split shifts of 7 A.M. to noon and 5 P.M. to 10 P.M. The industry argued the boys were starting a career in the skilled glassmaking trade. However, statistics don't bear that out. One industry estimate was that less than 4 percent of the boys would eventually be skilled laborers.[3] They had to work five years to be considered for an apprentice job.

In addition, boys were held a number of furnace-tending and general labor jobs. They were lehr (anneal) tenders, furnace tenders, batch-mixers, packers, and box-makers. Mike Owens had started his career as a ten-year-old furnace boy in a West Virginia glass house. General labor boys might work twelve hours, often through the night. Ohio ended the use of boys on the night shift in 1911 and restricted the number of hours worked to eight. Wages were from 30 to 50 cents a day. While a boy might make $3 to $6 a week, he might be charged $2 to $3 a week for room and board. Company boarding houses were common in Ohio. Orphans were often shipped from eastern cities to take the jobs in the glasshouses of Ohio where labor was in short supply. Companies hired boy-getters and might pay an orphanage or poorhouse for boys. This led to abuses such as the case in 1907 of a boy-getter kidnapping children in Philadelphia.[4]

Interestingly, Libbey used more women than most glass companies. The more agile and nimble fingers of women proved better at light bulb assembly and packing bulbs with straw in wooden barrels for shipment. Libbey also used women to sew covers for hand-blown products. Women made half what men made. In general, this meant fifty cents to one dollar a day for packing jobs.

Conditions for the boys at work and at the boardinghouse were often subpar. The union ignored boys in the wage demands, focusing on representing the skilled laborers. Boarding houses tended to be overcrowded, with 3 to 4 boys per bed. Some boarding houses used guards and dogs to stop runaways. The boys often had no shoes. The glass industry lobbied state houses to avoid any increase in minimum working age or improved working conditions. Owners often claimed the industry would collapse without child labor. Laws and inspections had less impact on child labor than automation and technology. To his credit, Libbey treated boys well. Boys did, at times, threaten strikes over conditions with some success. In fact, as we have seen, the initial motivation for process automation might well have started in Findlay, Ohio, in 1891, over a threatened mold boy strike. Children were also apt to run away, sometimes just for a day off. Tiffin Glass records note lost days and turns due to "Ringling's Circus," "County Fair," and "Christmas."[5] The boys may have been doing a man's job, but they were still boys. These interruptions resulted in major plant operations problems.

In general, Libbey Glass and the cut glass industry treated boys better. Libbey and his managers, Richardson and Robinson at Libbey Glass, had grown up in New England where the glass community tradition was entrenched. There, boys were brought into the trade as a career. Boys got valuable experience in the cut-glass industry, which could be considered a pre-apprenticeship. In fact, boys were much more likely to move into the skilled jobs in the cut glass industry. This career path was consistent with Libbey's interest in manual trades training. Boys in the cut-glass industry moved more rapidly into full apprenticeship. The Owens Bottle Company, on the other hand, used few boys because it had greater automation.

The first big reduction in child labor had come with Owens's foot pedal mold release that eliminated the need for a mold boy. The semi-automatic processes meant progressive elimination of boy laborers as productivity increased. Employment records suggest the semi-automatics may have led to a major reduction in child labor. Automatic machines eliminated most of the need for boys, but the number of machines in use grew slowly through 1919. The impact of the bottle machine on child labor was slow, as the early bottle machines eliminated blowers and gatherers rather than boys. However, improved productivity reduced the overall number of boys needed in the industry. A government study showed the increase in productivity from hand blowing to semi-automatic production: "In the 1860s the product of good workmen averaged from forty to forty-two dozen bottles of eight-ounce capacity per day. By 1905 the average output of the larger shop had become three hundred dozen per day, or 138 per cent increase."[6]

Some dispute whether the bottle machine had any effect between 1904 and 1913 because factories often used more boys to compete against automated plants. In 1904, child employment had fallen to 10.2 percent, with few bottle machines in operation,[7] probably because of higher productivity due to the semi-automatic process. By 1911, child employment had dropped to 4.7 percent.[8] A report in 1911 suggested that the early Owens machine represented a 33 percent improvement in productivity over semi-automatics. The report noted: "An average hand shop of a newer type, working on pint beer bottles, requires one boy to each to gross of bottles produced.[9] An Owens machine making the same ware requires one boy to each 15 gross of product." What is not disputed is that by 1919, with over 200 bottle machines making most of the bottles, the employment rate for children was 1.8 percent. It is clear that the automated process eliminated boys. In 1914, Toledo Glass announced the development of a fully automated anneal conveyor, which eliminated carry-in and carry-out boys. Automation eliminated the need for boys except for general labor.

Still, neither Libbey nor Owens had considered the machine as addressing child labor. In fact, Libbey's business strategy of licensing probably slowed the implementation of the labor-saving bottle machine. Had Libbey sold the machines on the open market, implementation would have been faster by years industry-wide. But Libbey realized that selling the machines would have destroyed the price structure of the industry and resulted in heavy loss of employment. In this he was not a humanitarian as much as a capitalist, with a kind heart for his employees and the community.

Libbey's income soared between 1905 and 1912. He took longer vacations in the winter months in California. In 1908, he and his wife first visited the Ojai Valley. He fell in love with this valley 85 miles north of Los Angeles and 30 miles southwest of Santa Barbara, even though his wife preferred the Pasadena area. (A lot of Toledo West Enders preferred Pasadena as well.) The Libbeys started to spend more time in the town of Nordhoff. The same year Libbey made some important purchases of paintings, including *Sheep on the Dunes* by Mauve ($58,000), *The Minstrel* by Frans Hals ($31,000), and *Amsterdam* by Jacob Maus ($12,500). Libbey also purchased *Lady with Shawl* by Pennsylvania painter Alice Patton, one of many he bought at the annual Carnegie Institute Art Exhibition. Libbey's early focus was seventeenth and eighteenth century masters of Holland (Rembrandt, Hals, and Vermeer) and Flanders (Rubens and Van Dyck). These artists were extremely popular in the Gilded Age and carried the highest prices. Libbey began to shop for the emerging paintings of the impressionists, where he could compete with collectors like J.P. Morgan and Henry Clay Frick. The highlight of the year for the Toledo Museum was a visit by future President Woodrow Wilson (then president of Princeton), who inspected the building of the new museum.

The museum also added some important collections in 1907 and 1908. One of these was a pre–Columbian collection of pottery given by A.L. Spitzer. Spitzer had purchased the collection from the Peabody Museum at Yale. This collection added to the museum's theme of manufacture and art, Libbey's vision. The Smithsonian said of the collection: "A careful study of the earthenware of this province leads to the conclusion that for America it represents a very high stage of development, and its history is therefore full of interest to the student of art."[10] A.L. Spitzer would make many donations of pre–Columbian pieces. His son, Carl, was a trustee and executive committee member. Carl Spitzer had become one of Libbey's closest friends. The Spitzers were among Toledo's wealthiest families, and were internationally known for their art collection. A.L. Spitzer had investments of real estate, banking, and railroads. He owned the Nicholas Building and Spitzer Building. Carl Spitzer got Libbey involved in Toledo real estate as a personal investment. He had been a member of the Tile Club and was considered one of the founding fathers of the museum.

The Spitzers and Libbeys often traveled together in Europe in search of art. George Ketcham added to the pre–Columbian exhibit with some cliff dweller pottery. Charles Ashley, friend and Libbey lawyer, donated *The Swimming Pool,* by C.C. Curran. Libbey donated a second painting by Toledo artist Edmund Osthaus, *A Good Sportsman*, after earlier donating *Afield*, by Osthaus. In addition, Toledo friends such as Charles Ashley and Robinson Locke made extensive donations to the museum's library. Libbey's sister, Sarah, then living in Brookline, Massachusetts, donated a cast of Chapu's *Joan of Arc.* The cast reproduction became popular with Catholic school children, as Joan of Arc had just been canonized.

By 1909, the Toledo Museum was getting national and international acclaim not only for its art, but its unique organization. The *New York Herald* noted:

> Special significance attaches to the new building which the Toledo Museum of Art is putting up. For in success of this museum directed by Mr. George W. Stevens, lies a lesson for every city anxious to have a public art institution but waiting for someone to come forward and endow it. The Toledo institution has no endowment, yet has forged ahead notwithstanding. The Toledo Art Museum was organized seven years ago and its work has been conducted without an endowment or maintenance fund other than yearly dues paid by members. The permanent collection is worth about $100,000 and has been donated piece by piece during the last few years.[11]

In London, museum trustee C. Spitzer made headlines when he purchased Romney's *Henry Richard Fox* for the museum for a price of $32,500.[12] Spitzer often had competition from collectors such as Henry Clay Frick and J.P. Morgan, who were also Romney collectors. Spitzer was well known in European art circles and his connection with the Toledo Museum brought international attention. Libbey also added one of his most important paintings, *The Mill and Castle,* by Hans Holbein.

In the fall of 1909, Libbey contributed to an exhibit at the Metropolitan Museum on great Dutch artists. The main focus was Rembrandt, and the great collectors and industrialists showed off their favorites. There were Rembrandt's *Portrait of Himself,* owned by Henry Clay Frick; *Gilder*, owned by H.O. Havemeyer; *Portrait of Himself,* owned by J.P. Morgan; and *Portrait of a Young Man*, owned by Libbey. Libbey also exhibited his *Boy Playing Flute* by Franz Hals. Other paintings from Morgan, Frick, Vanderbilt, and J. Gould filled the exhibit. Libbey had arrived on the national art scene, and kept an apartment in New York to be close. Over the years, he spent more time in New York to establish himself as a significant art collector. From 1909 to 1918, Libbey's art purchases made *The New York Times*, but his name was often incorrectly given as "Edward H." or "Durmente Libbey." Later in life, Libbey made some significant donations to the Metropolitan Museum to raise his profile.

Libbey's donations of art and funds to the Toledo Museum totaled over $200,000, far more than he had originally envisioned. His friend Carl Spitzer offered a proposal to the trustees to name the museum after him. Many have suggested that humility led Libbey to refuse, but his real motive was much more practical. He said, "The best interests of the Museum require that it should be universally felt that this is the People's Museum ... it having been found repeatedly in other cities that museums named after individuals who have made large gifts to them are ever after lacking in other donors and in that general support and enthusiasm which gives most life to such institutions."[13] Libbey realized that community involvement meant he could leverage his dollars in the acquisition of art, and build an internationally recognized collection. A "Libbey Museum of Art" would forever be dependent on him alone, and that could break him, or make the museum depend on government funding. Libbey's approach, a true community museum, was novel and creative, and proved visionary, as we see the result of tax-dependent museums and science centers today. Tile Club member and artist Thomas Parkhurst noted the democratic nature of the museum: "I have visited every museum of note in our country in the last few years, observing carefully their treasures and methods to gain the favor and goodwill of the public, and in none have I found such a democratic and helpful influence as our own."[14]

In 1911, the Libbeys made their usual trip to Europe, but it proved a difficult season. Most American tourists converged on Paris in July through early August. In 1911, a major cholera panic hit Paris, shortening the season. Libbey changed his plans in late July and went to Geneva by automobile for the last weeks of his stay. Still, he made some important purchases, including *Portrait of Lady Janet Trail*, an oil by Sir H. Raeburn for $105,000, the most expensive work Libbey had ever bought. But this purchase was overshadowed by the headlines when Henry Clay Frick paid $500,000 for a Thomas Gainsborough painting. Libbey added some other significant paintings in 1911, including Winston Homer's *Sunlit Shore* for $15,000; George Inness's *After a Spring Shower* for $6,000; and H.W. Ranger's *Landscape* for $2,255.

Libbey replaced his friend, Irving E. Macomber, on the Toledo City School Board in October of 1911. Macomber had pioneered using bond issues to develop city high schools and forged early ties between the school system and the museum. In

1912, Libbey was elected to the board, but his business commitments made it difficult for him to attend the monthly meetings. In his short tenure, Libbey reinforced the strong ties between the schools and museum, and established scholarships for teachers and students. Libbey resigned in 1914, but he continued his close relationship with the community schools, and donated to the schools and student scholarships over the years.

Libbey demonstrated outstanding people skills in all his endeavors, and started to build the museum's collection in preparation for the new building. Libbey was also active behind the scenes in Republican politics, which sapped some of his time. He was asked in 1912 to be a delegate for William Taft. Libbey had the strong backing of his friend, the Republican editor of *The Toledo Blade*, Robinson Locke (the *Toledo Bee* was the Democratic paper).

Libbey was active in the 1912 presidential campaign, which split the Republican Party between progressives and conservatives. The Ohio Republican delegation contended with the challenge of ex-president Theodore Roosevelt to then-President William Taft. Libbey and John Wileys led the effort to elect Taft. Libbey had the strong support of Toledo's wealthy, such as Robinson Locke, Arthur Secor, and Carl Spitzer, and stood nationally with Henry Clay Frick and other industrialists in their support of Taft. Businessmen were not sure where Roosevelt might go with trusts and tariffs. Libbey personally was a "McKinley Republican," supportive of protective tariffs and business growth, but Libbey, like most Americans, had embraced cleaning up corruption.

Libbey had supported reform mayor Brand Whitlock (1904–1914), who also was a museum trustee. Both Roosevelt and Taft campaigned in Toledo, and Roosevelt implied that Libbey had given money to local political bosses for favors and to obtain a delegate assignment. Libbey met the train when Taft came, and said in response to the Roosevelt attack:

> I have never been in favor of and never shall be in favor of any proposition to put up money to be elected to any public position. I am in this campaign at the personal solicitation of President Taft and because I believe in him. He has given this country a clean and honest administration and it would be a disgrace if the State of Ohio should not give him a solid delegation. I have no confidence in Roosevelt's word. I don't think the country can afford his leadership."[15]

This active political involvement was shorted-lived but frustrating. Taft won the nomination but lost to Woodrow Wilson.

Woodrow Wilson was a progressive Democrat, which local Republican industrialists feared the most. Wilson reduced tariffs, including those on the glass industry. The Underwood Tariff Bill of 1913 cut the glass tariff almost in half. Woodrow Wilson also established the income tax of 1 percent, which seemed so harmless at the time that many businesses even supported it. It was to be a tax on the rich only, limited to those making $500,000 or more. The tax was popular and ratified by the states as an amendment to the constitution in 1913. The presidential election was a rare instance of public politics for Libbey, who was happy to return to his own pursuits.

Libbey was interested in every detail of museum construction. He clearly wanted a building for the ages. The fundraising booklet noted:

> From the first the building committee, having in mind the growth of our city and the spacious site provided, determined to construct a building on so scientific a plan, with so artistic an appearance, that it would never, within any assignable time have to be torn down. Accordingly a style of elevation was adopted which has come down to us from Greece and which after 2,000 years, still maintains its primacy. The interior was made absolutely fireproof, and the arrangement was such that wings can be added to the building from time to time.[16]

During the building construction, planners decided to use marble instead of Indiana limestone, which had been originally suggested, at an additional cost of $10,000, through a special deal with the Vermont Marble Company. Arthur Secor gave the funds for the rich, beautiful and long-lasting marble.

It wasn't just Libbey and Stevens who were on fire for this new museum. Charles Ashley, Libbey's friend and lawyer, and a building committee member, showed the same passion. In a letter to architect Edward Green, Ashley stated the view of the committee: "I shall not consider this Museum a success unless it attracts the generality of citizens as much as the popular shows of the day. A museum ought not to be a cold-storage warehouse, but a place of true and real interest. This Museum aspires to be the headquarters of every club or society working for the higher intellectual and aesthetic development of Toledo and vicinity."[17] Ashley had been a Tile Club member with Libbey, and they clearly shared a similar vision.

But the museum of 1910 was much more than the new building under way. A Toledo editorial explained it this way:

> The Toledo Museum of Art is much more than a museum of art. . Few people realize how extensively the museum has entered into the intellectual, social and economic life of the city. Its influence permeates the school; it reaches out for the uplifting of countless homes; it enters the shops and factories, contributing to the education and directing the profitable pleasures of a great host of people to whom, consciously and unconsciously, it is an actual need; it is a great constructive element in our civilization, making daily better men and women, through them conferring its blessings upon the children — in a word, creating, year after year, better citizenship. No other institution in the city could take its place, if it were abandoned or suffered to lose prestige with the lapse of time. The schools, the churches, the press and all the local organizations to promote human progress have their special functions, but the art museum is alone in its great field of usefulness, exclusively and invaluably potent for the objects, which led to its establishment.... No other institution has more practical objects, and no other is more deserving the encouragement of practical men.[18]

Such praise is indicative of the local support Libbey and Stevens had created.

Libbey remained just as aggressive in expanding his glass companies for the future, using stock to motivate his managers. An admirer of Andrew Carnegie, Libbey copied his technique of giving stock bonuses to key employees. Mike Owens would have none of it, often selling his stock for cash. Owens gave the board of directors fits as he dumped his stock for cash or gave it away. Libbey, however, continued to try to improve Owens's managerial and social skills. Owens did move into the West End, buying a thirty-room mansion at 2345 Collingwood near the Libbey home. But Owens and Libbey never became social friends. Owens and his wife lived quietly in the neighborhood, avoiding major social events, except neighborhood games. He preferred his home library of history books, having no taste for art or paintings and caring little about the museum. Libbey tried in vain to interest Owens in paintings. Owens loved to spend money freely, which seemed to annoy the company directors. He had fine clothes and several electric cars, which he preferred to the gasoline models. He was active in the Catholic Church and Irish national organizations. To his credit, he gave generously to church, family, and friends. If someone in the community needed help, Owens would give.

Chapter 14

Grand Opening

The year 1912 was momentous for Libbey. The museum would open its doors, and Mike Owens and Toledo Glass would launch a new project for the last glass industry segment untouched by the Libbey empire. This was unexpected; most thought that Libbey, at 56, would quietly retire to California and focus on art collecting trips. But Libbey kept his research company on track, ever improving the Owens bottle machine. Toledo Glass produced almost 300 versions from 1908 to 1927.[1] Libbey pioneered machine standardization with his engineer Emil Bock. Libbey had created a developmental dynamo in Toledo Glass, and for managers like Mike Owens, there were more horizons to conquer. But Libbey's main focus would remain the new museum, which was to be his legacy, although he refused to have it named for him. Libbey and his wife would expand their travel to support their art collecting. They started to spend more time in Europe and New York.

The museum building had its final inspection in late 1911 by the building committee members: Edward Libbey, Charles Ashley, Arthur Secor, Aaron Chesbrough, and George Stevens. Actual expenses had run over $300,000, and an additional fund raising campaign took place in 1911. The costs included a new home for Mr. and Mrs. Stevens. The museum raised another $100,000 and Libbey added $25,000. The finance committee and I.E. Knisely of Northern National Bank did an excellent job. Well over 600 people contributed to the building fund, and the museum had over 1,000 paid memberships. Libbey and the museum trustees were shrewd custodians, considering that the similar but less beautiful Buffalo Albright Gallery cost $700,000 to build. Libbey's leadership was every bit as focused as his leadership in his glass companies. Of course, Libbey was blessed with two creative work horses in Mike Owens and George Stevens. The trustees at the time of the grand opening were Charles Ashley, Clarence Brown, Aaron Chesbrough, John Doyle, Edward Ford, Fredrick Geddes, William Hardee, Frank King, Isaac Knisley, Albion Lang, Edward Libbey, Jefferson Robinson, Charles Schmettau, Arthur Secor, Fredrick Shoemaker, Barton Smith, Carl Spitzer, Irving Squire, David Stine, Alvin Tillinchast, and William Walding.

As the museum's grand opening approached, the Libbeys rushed to get art for the spacious shelves and galleries. They added over 1,400 items alone in 1912, among them Rodin's *The Thinker*. In December 1912, a month before the opening, the Libbeys were in New York lining up additional loans. In this last minute effort, Libbey purchased Raeburn's *Lady Janet Traill* and Winslow Homer's *Sunlight on the Coast.* Florence Libbey established the Maurice A. Scott Gallery in 1912 in memory of her father. This collection highlighted American painters, and included Frederic Remington's *Indians Simulating Buffalo,* Gilbert Stuart's *John Ashley*, and Childe Hassam's *Summer Sea.* She also donated an extensive ceramic collection, and personally took charge of cleaning the pieces for exhibition. They included extensive pieces by Josiah Wedgwood. Libbey added 40 drawings and watercolors by John La Farge. In an effort to extend the scope of the museum, Libbey added an array of textiles from all over the world, stemming from his early interest in textiles. They included embroidery, damask, and brocade from Europe and Asia. Libbey also wanted to expand into the technology of art. He purchased a mixed collection of Japanese and Chinese items from Gunsaulus for $3,490, including 120 sword guards known as Tsubas and several sword sets. Gunsaulus played a major role in the development of the museum and Libbey as an art collector, interesting Libbey in Asian art. Gunsaulus lent a seventy-five-piece collection of classical, Venetian and German glass for the opening, and donated Henry Tanner's *The Disciples on the Sea*, and Jose Weiss's *September on the Arun.* Gunsaulus even added rare books to the extensive museum collections.

The grand and lavish opening of the museum on Wednesday, January 17, 1912, was an event that Libbey had been anticipating for years. The day was cold and snowy, holding down the size of the crowds. The scope of the building alone placed it in the ranks of the New York Metropolitan, Museum of Fine Arts in Boston, the Chicago Art Institute, and Pittsburgh's Carnegie Institute. In fact, it was amazing that the thirty-fourth largest city in the United States had the fifth largest art museum. The opening included a memorial address by Dr. Gunsaulus. A special luncheon was held at the Secor Hotel for guest speakers. The luncheon menu included cream of tomato Bellevue, oysters Poulette, assorted sandwiches, lobster salad, Neapolitan ice cream, and Secor coffee.

Local patrons and special guests remained for events on the next day. A midnight dinner was given in the Gold Room of the Secor Hotel. The dinner included music by the Toledo Symphony. The midnight dinner would long be remembered as being in the tradition of the Scott family, who were fond of entertaining. The menu included wines such as a 1900 Montebello (1900), gumbo of volaille, viande of crab au gratin, poussin grille (with petit pois and pomme Parisienne), salade de pample mousse, biscuit tortoni, fromage de Philadelphia, and Apollinaris bottled water. Cigars and cigarettes were offered to the men after dinner.

Gunsaulus lined up loans of several collections of Dutch masters, including a collection of works by Dutch painter Josef Israels, a friend of both Libbey and Gunsaulus. Israels loaned one of the paintings to Libbey, a self-portrait inscribed "for Mr. Libbey 28 Oct. 1908." Libbey and Gunsaulus had held a dinner in The Hague for Josef Israels at which the painting was presented to them. *The New York Times* highlighted the following:

> The large western gallery is filled with examples of the great musters, notably three superb Gainsborough portraits and a Romney double portrait, loaned by Henry E. Huntington of New York; Henry C. Frick's El Greco's *Saint Jerome*; a Romney portrait loaned by Mr. Ford of Detroit; a Hals and Rembrandt loaned by Mr. Libbey; three Mauves from A.S. Stephens of Moline, Ill., Mrs. Tobbey of Detroit, and numerous other superb examples of Baltzer and other modern Dutch masters. There were a number of loaned works of James Whistler, who was extremely popular at the time. In the east and small galleries were 112 examples of the strongest modern American painters, from Alexander to Wyant, comprising the best works shown the public in past years in galleries of Eastern cities. One panel of nine Whistlers was loaned by Messrs. Freer and Canfield.[2]

The first day included a presentation to the founders, which included gifts for Libbey and Stevens. Libbey's lawyer, friend, and building committee chair Charles Ashley officially turned the program over to Libbey, its president. Libbey accepted with the following:

> This beautiful building which we are to dedicate today, is the embodiment of an impulse of far greater magnitude than this our local manifestation would suggest, for it is in truth one of the many evidences of a great national awakening; of the lofty present-day aspirations of the American people and of the refining influence of a mature and perfect civilization.... Today in this community teeming with its manifold activities the necessity for the practical and economical application of art was never more apparent. We can never be classed as a great city if we deprive ourselves of this source of mental and spiritual stimulus so necessary in forming the character of our people and instilling in them those ennobling and higher ideals of life.[3]

The themes of other speakers reflected Libbey's vision of art, community, and industry. Dr. Frank Gunsaulus's speech was titled "An Art Museum in a Manufacturing City," and the Rev. George Wallace of First Congregational Church preached on "The Message of Art and Its Relation to Religion." Additionally, Charles L. Hutchinson, president of the Art Institute of Chicago, talked of the role of the museum in setting an ideal for the community. Gunsaulus's speech best reflected Libbey's vision of the museum, which in turn was deeply affected by Gunsaulus's philosophy of art.

The poor weather continued on January 18 as the museum opened to the public. The inaugural exhibition would not disappoint the 1,500 who braved the weather and formed a long line. The exhibition included over four hundred paintings, some owned by the museum, some loaned for the event. There was an outstanding memorial exhibition on Josef Israels put together by Libbey and Gunsaulus. The National Sculpture Society put a special exhibition together. The Egyptian exhibition continued to be a favorite of the locals. The newly printed catalogue noted that the mummy's coffin was "opened, disclosing that which has been partly unwrapped to reveal the remarkable preservation of the body which has been in its present condition for over twenty centuries." The inaugural exhibition eventually drew over 50,000. One of the sculpture busts was of President William Taft, who would visit a few weeks later.

After the opening, a school of art was established directed by Almon Whiting, who had been the first curator of the museum. Whiting was an established artist who had studied at the famous Julian Academy. The school of design would be the start of a bigger and broader mission for the museum. New school trustees included Edward Libbey, Frederick Geddes, Carl Spitzer, Charles Schmettau, Brand Whitlock, John Willys, and Irving Macomber, the core of Toledo's businessmen. Secondary committee members included George Stevens, Robinson Locke, William Gosline, Charles Ashley, Arthur Secor, and Clarence Brown. Thanks to these men, the museum was debt-free. The art selection committee consisted of Libbey, Arthur Secor, and George Stevens. Robinson Locke chaired the library committee. George Stevens remained director, and Almon Whiting was named curator. Officers were Edward Libbey, president; William Hardee, first vice president; Arthur Secor, second vice president; Charles Schmettau, secretary; and Isaac Knisely, treasurer. The trustees now included two more important Toledo industrialists, John Willys of Overland Automobile and Edward Ford of Ford Glass. John Willys was a major collector of Dutch artists such as Rembrandt, and his home in the West End was known for its beautiful paintings. First Vice President William Hardee was a new officer Libbey sought for his managerial and executive skills. Hardee was a nationally known businessman who had formed National Supply Company. At the time he was president of the Ohio Savings Bank & Trust Company. He had previously made a fortune in oil, gas, and banking and was a member of many business clubs in Toledo, Pittsburgh, and New York.

The new board addressed the need for maintenance and security personnel as well as staff professionals. The librarian was given an assistant. A building manager and assistant were hired. Fees were increased: sustaining memberships went from $50 to $500, and included a core of wealthy business patrons. Life membership was established at $1,000, and fellowship membership at $5,000. The annual $10 membership remained, but a new associate membership was established for students, craftsmen, and teachers. With the museum opening, memberships soared. The museum's public support became the envy of other cities.

The inaugural exhibition was hailed across the nation. Charles Freer of the National Gallery noted: "This is the most remarkable collection ever shown in this country. As yet you have no idea of the valuable asset you have in your museum. It places Toledo in the foremost ranks of American cities and at the head of the artistic development of the great middle west."[4] The *American Art News* went even further in its praise, with: "The inaugural exhibition ranks first among any similar displays in the fact that no more representative showing of modern American art, combined with the representative display of early modern and foreign art, has yet been made in this country."[5] Maybe the most important praise came from Theodore Keane, director of the Minneapolis Society of Fine Arts: "What was most noticeable in Toledo was the art spirit of the man on the street. Every one talked art and seemed to have appreciative views regarding the movement which resulted in the erection of the splendid museum recently opened."[6] What most art critics missed at the time was a shift in theme. The museum approached art in very broad terms and views. It made a link between art and manufacture not explored at other American museums, but similarly explored in Germany. Libbey, with his glass manufacturing background, influenced the museum in this direction.

The Toledo Museum of Art evolved to a large degree from 1901 to 1925. The mission in 1901 was to:

> establish and maintain a gallery for the exhibition of paintings, sculpture and other works of art and a museum of natural and other curiosities and specimens of art and nature, promotive of knowledge, and to establish and maintain an academy for advancing, improving and promoting painting, sculpture, drawing, architecture, and other refined arts and furnishing instruction therein by lectures and otherwise, at the City of Toledo, in the State of Ohio. All to be open to the public on reasonable terms.[7]

This was a broad theme which remains to this day. In 1914, the museum joined the "City Beautiful" campaign, which included a two-room exhibit on home decor. The museum added a bird study program for children, which attracted a donation of a statue of naturalist John Burroughs from Libbey engineer and inventor Emil Bock. Stevens even started some boys clubs on collecting coins, stamps, and arrowheads. There is no doubt the museum struggled with direction. Stevens, Libbey, Florence Libbey, and Gunsaulus collected a broad array of art. The early museum reflected the early collecting habits of those four. Eventually, Libbey would tie the diverse directions of the collections into a vision of manufacture and art.

In 1913, Libbey purchased an extensive collection of ancient glass, which would be the start of one of the world's best glass collections, from Julius Campe of Hamburg, Germany. Gunsaulus set up the purchase. The Campe collection was a study in the history of European glass, from Islamic glass to Venetian glass to Northern European styles. The collection formed the historical core of the museum's glass collection. The Humpen cylindrical beakers and Roemer goblets were part of this early collection, which included Waldglas, produced in a German forest. There were also examples of Latticinio and Cristallo glass. Gunsaulus was a glass collector long before Libbey and had made donations to the museum. This collection was added to a collection acquired from George Olcott of Columbia University. The Olcott collection covered early

Syrian glass of the fifth century B.C. to the second century A.D. The museum trustees added six pieces of early Egyptian glass, including cast jars dating to 800 B.C. Libbey added some other ancient pieces such as a Phoenician blue bowl and a Tyrian purple goblet. The year 1913 set a record in attendance, at 97,234. This represents 57 percent of the Toledo population, outpacing all other museums based on city population.

Chapter 15

The Story of Flat Glass

The Owens bottle machine brought the first major change in glassmaking in 3,000 years, stirring a spirit of innovation throughout the industry. At the start of the twentieth century, the glass industry could be divided into four segments: plate glass, window glass, glass containers, and pressed or blown glass. Window and plate glass manufacturing represented 34 percent of the industry in 1904. The growth of department stores and office buildings was created a boom in plate and window glass throughout the nation.

Window glass technology had progressed little from early times. It was produced by an old, complex method of hand manufacture starting from blown cylinders. It was a blown glass product made from blowing large cylinders, then cutting and flattening the glass to sheets. Cylinder blowing was the highly skilled purview of European glassmakers. Many Belgian glassmakers immigrated to Ohio to make cylinder glass. The Owens semi-automatic approach to bulb and bottle making inspired similar applications to compressed air blowing, but this process was slow to replace the method of the Belgian artists. The revolution in flat glass would again involve Libbey and Owens. It would be a challenge on many fronts. The biggest challenge would be from the most powerful union in the history of the United States — the Local Assembly (Knights of Labor) 300.[1] Window glass production made up over 50 percent of industry employment. The technical challenges were almost as great as those posed by the union, but Toledo Glass had the management and talent to overcome the obstacles.

The two traditional means for making window glass were the crown glass method and the cylinder glass method. Crown glass was popular for making small window squares from colonial times. It was produced when the blower made a large globe and transferred it from his blowpipe to a pontil rod. The rod was then swirled by an assistant until the globe collapsed into a flat round plate on the end of the rod. The plate was cracked off the pontil rod, creating a bull's eye scar and cut into square panes. The bull's eye can be seen in glass from old colonial homes. Crown glass was limited in the size that could be swirled by hand. The clarity of the product was poor, with a distortion caused when the plate formed by collapsing the globe.

The cylinder process was extremely labor intensive. First, 20 to 60 pounds of glass was gathered on a blowpipe which, because of the weight, was rested on a large wooden block for blowing. The glass blowing was done in a trench and required frequent reheating to produce a large cylinder. The blower might swing the cylinder to lengthen it. The final cylinder was from 10 to 12 feet in length. The process required great skill and strong arms and lungs. After blowing, the cylinder required annealing. The ends were cut off to create a tube and it was slit down its length. Finally, the tube was flattened into a piece of glass through reheating and annealing. The flattened sheets were taken to the cutting room. Unfortunately, even this method created some rippling and distortion. Cylinder glass took many man-hours to produce.

The labor-intensive cylinder glass industry was located mostly in Pittsburgh after the Civil War. It was a booming industry throughout the 1800s with the Republican tariffs in place. In 1867, Pittsburgh window glass blowers organized one of the first glass industry unions, part of the trade labor movement of the Knights of Labor. In 1878, the union inducted window glass gatherers, flatteners, and cutters. This consolidation resulted in Local Assembly 300, Knights of Labor, Window Glass Workers of America. With the Knights' national ties, the window glass industry quickly became nationally unionized. Local Assembly 300 standardized procedures on the ancient Venetian crafts system. By 1881, Local Assembly 300 in Pittsburgh's Monongahela River Valley controlled labor at all of America's window glasshouses. The union set daily wages at $1 a day for blowers, 57 cents for gatherers, 27 cents for flatteners, and 27 cents for cutters, splitters, and cappers, all percentages of the blower's $1. Local Assembly 300 would evolve into one of America's most powerful unions, winning many pay increases and building a wage structure in the guild organizations it represented throughout the 1870s and 1880s.

The union also won restrictive work rules. These limited the numbers of cylinders blown per hour. Flatteners were limited to four pots of glass and cutters to two and a half pots. Work ran from September 1 to June 30. Only sons and brothers of union workers could become apprentices and lasted three years. Members were banned from working with automated equipment. This powerful structure was reminiscent of the old guilds of Europe. In 1895, new technology saw the rise of automated blowing in the cylinder glass industry. At the start of the twentieth century, Local Assembly 300 still was resisting new technology, despite rapidly changing conditions. The union split with the Window Workers of America over internal funding, which helped spread automated blown cylinder technology. Around the same time, window glass manufacturers formed a monopoly in 1895 as a unified front against the national union. This company, American Window Glass Company, included 53 glasshouses and 70 percent of the manufacturing capacity.

Initially, American Window Glass Company cooperated with Local Assembly 300 and slowed the use of automation. When continuous tanks for melting became available instead of the furnace pot, American Window held back to some degree. American Window's control of the market forced the independent companies to unite in a selling pool, American Window and the unified pool of independent producers controlling the national market. The price of window glass soared. Eventually, the government would win anti-trust cases against the industry. Short term, the union even joined in the monopolistic structure. The companies promoted Local Assembly officers into management and offered wage increases for changes in work restrictions.

By 1903, an automated blowing machine invented by John Lubbers allowed for the blowing of huge cylinders. Lubbers had the financial support of James Chambers, president of American Window, who purchased the patent rights. The Lubbers machine could produce a 50-foot-long, 30-inch-diameter cylinder, compared to the 6-foot, 14-inch cylinder produced by a hand blower. The machine offered large gains in productivity and eliminated blowers and much of the handling. A flat glass plant was built in Maumee, Ohio, using impressive Lubbers process.

The union control and pricing arrangements could not hold back the advance of technology. Inventors were working on a way to cast glass sheets directly. In France, Emile Fourcault was building a vertical drawing machine; and in the United States, Irving Colburn was working on a horizontal drawing machine. In 1914, the Fourcault process was still not fully operational. Toledo Glass actually employed Colburn's brother and was well aware of his progress. Colburn's experiments were so successful by 1906 that *The Toledo Blade* announced: "The glass industry of the United States and the world is in another furore similar to that caused by the invention of the Owens machine for making bottles automatically, and this time also it is a Toledo man whose genius has produced a wonderful result."[2] The newspaper had adopted Irving Colburn as a native son, but its praise was premature. The potential, however, could not be overlooked. The machine could reduce labor costs by 85 percent. For the next four years, Colburn struggled with financing commercial experiments in his Pennsylvania factory.

In 1910, Mike Owens and several Toledo Glass engineers visited the Colburn operation to look at its potential. The process had technical difficulties. Owens and the engineers reported back to the Toledo Glass board that Colburn's design had potential and the company should consider investing. The board balked, noting that Pittsburgh Plate Glass had turned Colburn down. Lawyers Geddes and Brown were concerned about possible lawsuits. Toledo Glass lawyers estimated that Colburn could encounter over 30 patent infringements, and knew that more than twenty glass companies had refused to invest. Walbridge hesitated because of his concern over costs and his desire to further improve the bottle machine. Libbey had personal reasons for not wanting to launch a major project. He was president of three corporate boards and several community boards, and the museum was to be opened in 1912. In addition, Libbey was making annual business trips to Europe, Florida, and California, and

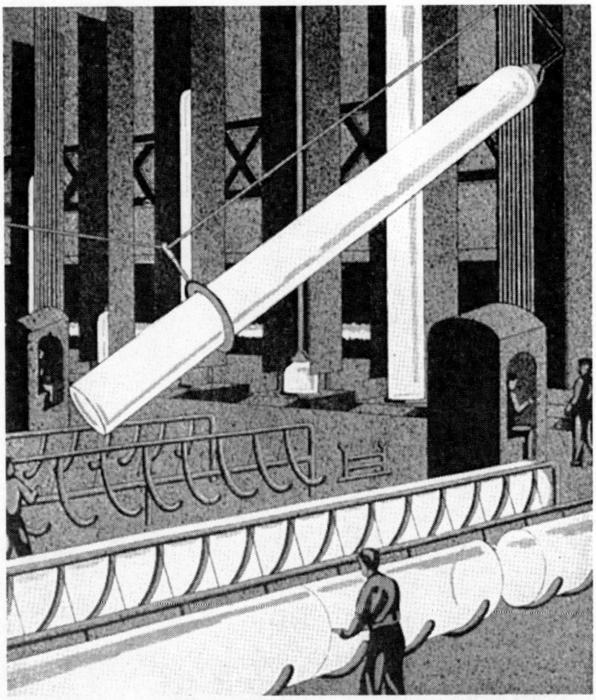

Lubbers process for large cylinder glass production (courtesy of University of Toledo, Ward Canaday Center, Libbey Glass Records).

often made collecting trips to New York. Florence Libbey had hoped that her husband, now 57, might consider a retirement of travel and art collecting. Owens remained the sole supporter of investment in Colburn's work from 1910 to 1912, but Libbey had far from closed the door on Colburn. Libbey had initially heard of the design in 1908 in a letter from Colburn's father, who lived in Toledo. More than anyone, Libbey understood the investment potential of a flat glass machine. He refused to rule out the possibility, allowing Owens to pursue the idea and a relationship with Colburn.

Mike Owens continued a relationship with Colburn, probably with the tacit support of Libbey. As we have seen, Libbey's perceived indecision often worked to his advantage. James Chambers, president of American Window, never closed the door to Colburn either. Owens and Colburn became close friends, playing golf often at Inverness Country Club. The two visited other glass plants to discuss technology. In late 1911, Colburn's company went bankrupt, and the assets were to be sold at auction. Owens threatened to leave the Libbey fold and start his own company to develop the Colburn machine, and brought Libbey to a decision. Owens got Libbey's approval to purchase the Colburn machine for $15,000 in February of 1912. Experts believed another $125,000 might make the machine operable. Libbey added Colburn to the Toledo Glass payroll and the project was underway. Colburn was set up with a new assistant, Joe Crowley of Libbey Glass, and an engineer, Libbey's first cousin, Arthur Fowle. Toledo Glass had a number of active projects in bottle production. Libbey hoped that his brother-in-law, the general manager of Toledo Glass, would keep the costs in line. Walbridge liked Owens, but he often would say, "He'll experiment us to death." Still, Walbridge had learned to respect Owens for his successes.

Libbey knew Mike Owens better than anyone, and he knew what was possible. Regardless of any threat to leave, Libbey agreed with Owens once he was convinced of a project's potential for success. To many, it was a strange relationship, but it was one of the most successful in the annals of American business. Libbey had built an entire organization around Owens to handle his technical and managerial weaknesses. One writer noted:

> Libbey and Owens made an excellent, harmonious team for what they originally set out to do. Everyone who knew Owens agrees that he was intractable and could not buck supervision which interfered with his work. This is scarcely surprising, for genius usually is intractable and withers away or becomes frustrated if subjected to controls ineptly administered. Libbey was one of the few people who understood Mike and had the ability to direct his energies into their most fruitful channels. Libbey patiently supplied the necessary funds and encouragement.[3]

It is clear that Owens's commercial inventiveness depended on Libbey, much as Tesla depended on Westinghouse. Libbey and Westinghouse built America's first research and development operations. It would be hard to recognize the genius of Owens without acknowledging that of Libbey.

The board came around with Libbey's support and an experimental plant was approved for a 6.5-acre lot on Toledo's Castle Boulevard. The flat glass machine, known as the CX machine, was to be working experimentally by late 1913. Walbridge reported to the board in November 1903 that total costs had reached $252,819. This was a large amount, but Toledo Glass had the cash, and the potential was huge for this machine. While some see Toledo Glass as Owens's company, the truth is that Libbey and the board had built an army of technicians and engineers. There was no similar development company in America at the time. Libbey had set up its funding based on its own returns. Toledo Glass operated at a surplus during the Colburn machine development period but did suspend dividends for a few years. Toledo Glass was held in high regard. The industry noted in 1912: "The reputation of the Toledo people for carrying to a successful conclusion whatever they undertake is pretty generally understood, and it may be taken for granted that they would not now be backing the sheet drawing patents if they were not convinced of ultimate success."[4] This was true of Libbey. He often wrestled with whether to take on a project, but he rarely let go if he smelled success, witness his move to Toledo, and his investments in the World Fair and the bottle machine.

While the flat glass machine development struggled on, Libbey's key managers were being sucked in to the huge project. W.S. Walbridge was functioning as treasurer for Toledo Glass and vice president and treasurer of Owens Bottle. Libbey had told the board of Toledo Glass that he would create a flat glass company, once the CX machine was commercially viable. Walbridge and Libbey were active in the Commerce Club, which had a young University of Michigan graduate, John D. Biggers, as secretary. Biggers was asked to be assistant treasurer of Owens Bottle Company. Biggers was happy to move into the Libbey organization but had to take a pay cut from

$5,000 a year to $4,000 a year. John D. Biggers would be a key second-generation executive for the Libbey empire, and his presence freed up Walbridge for the flat glass business.

The first success of the CX machine came on Thanksgiving 1913, while Libbey was in California and Owens in Florida. Owens tried to play Libbey's role by encouraging Colburn and his team. The product that Thanksgiving was the most promising since Toledo Glass took over development. The problem remained with the roller system needed to pull out the horizontal sheet of glass. The rolls marked the glass and a search started for roll materials. The best minds of Toledo Glass were at work, but it took another two years of work, which tried the patience of the Toledo Glass board. Materials such as clay, asbestos, and Monel (nickel-copper) were used. The solution was a cutting edge metal, Nichrome. While work continued on the roll materials, engineers upgraded the whole process. When Owens returned from Florida and Libbey from California, they were extremely pleased; but a lot of work was still needed.

Just after the Thanksgiving success, Toledo Glass treasurer William S. Walbridge estimated that costs had already run over a quarter of a million dollars and much more would be needed. The board approved an additional $100,000. A few weeks later another $30,000 was requested. The next two years would push the Toledo Glass board to the limit. By the end of 1915, Toledo Glass had cash flow problems. Owens kept his reports extremely positive while facing one technical problem after another. In particular, the sheet glass had edge knurls, which required re-work to produce a usable product. Owens proved invaluable during this period. By edge trimming, the CX machine was producing usable product in the spring of 1915. The first customer was Smith & Wyman Sash & Door Company. Development costs were approaching a half million dollars. The once-supportive industry news sources believed Libbey had failed, and for months had been giving gloomy reports on the status and financial condition of the project. Again behind the scenes, Libbey refused to fold. Libbey sold the Owens European bottle operation for over $80,000 in 1915, which brought cash in for the CX machine project. As with the bottle machine, without Libbey the board would have ceased development, as would most in the glass industry. Owens and Colburn remained confident, however, through all the problems.

By fall 1915, Libbey deemed the project near enough to success to look to the future, as they had with the bottle machine. Libbey again would prove the organizational visionary. The board and Libbey, not willing to put Owens in charge of plant site selection, turned to Libbey's cousin, Arthur Fowle, to do the site evaluation. Charleston, West Virginia (actually Kanawha City), was the site chosen. Libbey teamed up with his Ohio Bottle partner, Edward Everett, who had ties in West Virginia, to assure a natural gas supply. In early spring, the CX machine was in commercial operation. By May of 1916, Owens reported to the board that all the problems had been resolved. He was a bit optimistic, but the board moved for the establishment of Libbey-Owens Sheet Glass Company for commercial development of the machine. The results lived up to expectations, with the experimental plant drawing sheets 48 to 68 inches wide at a speed of 36 to 60 inches per minute. It was all done with a machine operator, saving 90 percent on labor cost.

The immediate problem of the Libbey-Owens Company was what to do about Mike Owens. He did not fit into normal organizational structures, and even Libbey realized it. Unlike the board, however, Libbey was sensitive to Owens. Libbey was in California, but the issue had to be resolved. Clarence Brown pushed the matter with Libbey, and Libbey initially responded: "I believe that with proper restrictions, properly prescribed, Mr. Owens would make the ideal manager and, as he is really the only available person at the present time, I'm inclined to recommend his appointment to that position."[5] That was in February 1916, and the board still hesitated. As with the bottle machine, Libbey wanted to set up a new company to manage the CX machine. On May 18, 1916, Libbey-Owens Sheet Glass Company was established. The company capitalization was $6 million, with the incorporators being Libbey, William Walbridge (treasurer of Toledo Glass), Fred Geddes (secretary of Toledo Glass), Clarence Brown (vice president of Owens Bottle), and Charles Schmettau from the Geddes and Brown law firm. Libbey-Owens paid Toledo Glass $4 million for machine rights in common stock. The remaining common stock was divided among the incorporators. Libbey-Owens sold $1.5 million in preferred stock to raise cash and later sold the experimental Toledo plant to Willis-Overland Automobile for $165,000. Once again, Libbey raised the investment capital without sharing control of the new company. Both Owens and Colburn received sizable amounts of stock for their work.

The decision over whether to name Owens general manager dragged on. Brown finally realized that he had no choice but to back Mike Owens. The board was very concerned about putting Owens in a major operations position. Clarence Brown sent the following to Libbey at Ojai:

> Concerning the temporary organization of the Libbey-Owens Company, the situation is this. It had been suggested that I should be named President and Fowle vice-president, while Owens was not even to be made a director. It occurred to me that this would not be wise and certainly not agreeable to Owens. There seemed to be no sufficient reason why Fowle or I should be put, even in temporary authority over Owens, nor any justification for not putting Owens on the Board of Directors, in view of his past connections with the enterprise and his large financial interest in the future of the Company.... I think Owens's cooperation in the enterprise is very important and I have no doubt, when it comes to a permanent organization, you will agree with me that he should be placed in a position of authority.[6]

The new officers were Libbey, president; Clarence Brown, vice president; Charles Schmettau, secretary; and Arthur Fowle, treasurer. Owens and Walbridge were added to the board of directors at Libbey's request. Owens was named general manager, and Bill Jacobs was named manager of the new plant. Jacobs had worked on the development team as Colburn's assistant and was a friend of Owens. Arthur Fowle was in charge of the new plant's construction. Colburn's development crew was given the opportunity to purchase stock at par value. The market rewarded all as the stock rose from $25 to $40 a share. In dollars terms, stock bonuses for Owens, Colburn, and Jacobs amounted to $750,000. There was another $250,000 for stock bonuses throughout the organization. Libbey had become convinced, like Andrew Carnegie, that stock bonuses were motivational. Owens was quick to convert to cash. Libbey, Owens, and Colburn all set out on different long vacations.

On his return, Owens lost no time in aggravating the board of Toledo Glass by putting forward a motion to dissolve Toledo Glass and distribute the stock. Owens wanted cash; he had little interest in stock. The motion was defeated with only Owens voting in favor. He resigned as a director of Toledo Glass. It was January and Libbey was at his Ojai home, but suggested a payment for past services to Owens. The board agreed to pay Owens $200,000, and he sold additional stock. The board also agreed to pay Owens a one-cent royalty on each gross of bottles made on his machines.

Libbey wanted to turn Owens's focus toward the growing competition to Owens Bottle Company and the Owens machine. The Graham Glass Company of Indiana had developed a new bottle-making machine which was running successfully at its Evansville, Indiana, plant and was attracting a lot of attention. Graham had converted a semi-automatic process into an automatic one. The new process was a gravity fed mold system. Gravity feed eliminated the gatherer. The Graham machine cost about a fourth of the Owens machine and was easy to install and more flexible at producing small orders. The Graham machine merely had potential in late 1915, but it was a major concern to Libbey, who tried unsuccessfully to buy the patents. Libbey instead bought Graham Glass Company on June 28, 1916, with Owens Bottle Company taking ownership. Libbey realized that others were developing gravity fed machines, and he directed Owens to improve the gravity feeding system and try to incorporate it in the Owens machine. This was where Owens's talent would best be applied.

Libbey was one of the few in the company to fully realize the threat of gravity fed machines. In 1916, Owens Bottle had only 50 percent of the market, and a cheaper process could undercut even that proportion of market share. Graham was only one new competitor. The Hartford-Fairmont Company was working with Beechnut Packing Company to develop a gravity fed-type machine. Libbey again tried to buy all competing patents but was not completely successful. He even filed a patent infringement suit against Hartford to slow market penetration. In the meantime, he urged Owens to improve the bottle machine. This development work was done at both Owens Bottle and Toledo Glass. Owens's ongoing work at Owens Bottle, Toledo Glass, and Libbey-Owens Glass left both him and the companies without focus.

Libbey's attention was being torn among development projects at a number of his companies.

The Toledo Glass CX machine and the improved bottle machine were only two development projects Libbey was involved in. At Libbey Glass, the board had incorporated Westlake Machine Company to engineer an automated light bulb machine. Westlake, organized in 1907, functioned as a research arm of Libbey Glass, completely separate from Toledo Glass. The automatic bulb machine created the same passion at Libbey Glass as other projects at Toledo Glass. Libbey, Richardson, Robinson, and Donovan (all from Libbey Glass), as well as Owens and Walbridge were the major stockholders of Westlake Machine. Richardson was pushing the automated bulb-making machine. Cornelius J. Nolan, who had a number of patents, managed the experimental shop on Westlake Street in Toledo. Nolan was a skilled engineer but Westlake lacked the organization of Toledo Glass. Work on the Westlake automatic bulb machine continued while Toledo Glass worked on the CX machine. Richardson was the mainstay at Westlake, as Libbey's time was divided. Libbey remained president of Toledo Glass, Libbey Glass, and Owens Bottle.

Libbey's work schedule was heavy, but he retained a great deal of freedom. His managers included Richardson at Libbey Glass; Brown, Walbridge and Owens at Toledo Glass; and Brown and Geddes at Toledo Glass and Owens Bottle. At the museum, Libbey had George Stevens. Libbey had built a very strong management structure throughout his organizations, which allowed him time to pursue his passions. Besides Libbey, the museum had the executive skills of Richardson, Geddes, Charles Schmettau, and Clarence Brown. Interestingly, while he was a sustaining member ($50 a year) of the museum, Owens had little interest or belief in a community art museum. Owens could not understand collecting art. Clearly, Libbey's closest business associate was Clarence Brown, who besides being a friend and lawyer, served on most of Libbey's boards, with the exception of Libbey Glass. Brown worked often behind the scenes but was a loyal lieutenant. He was probably the second biggest benefactor from Libbey's business success.

Art collecting and travel took up a substantial amount of Libbey's time. In 1908, Libbey had been introduced to the small California town of Nordhoff (by 1917 known as Ojai). The Ojai Valley's air was thought by many to have restorative powers. Libbey saw the small town as a peaceful, rustic getaway. (Interestingly, Ojai was used as Shangri-La in the 1937 movie *Lost Horizon*). Florence Libbey found little that was charming in the small town, preferring the suburbs of Los Angeles, but Libbey was convinced it would be his personal paradise. In 1909, Libbey contacted Los Angeles architect, Myron Hunt, about building a craftsman bungalow. "Palace" would be a much better description, but many additions were made over the years. The original home on Foothill Road had additional cottages for the servants. The home had a beautiful view of the valley, and Libbey loved to ride his horses there.

By 1914, Libbey set about redesigning the little community. He had been interested in real estate development in Toledo, but he found the city too big to fully influence. Much of his real estate ownership in Toledo was in commercial properties such as the Valentine Building, the Lasalle Building, the Colonade Building, and much of the commercial district on Toledo's Adams Street, all revealed after his death. He envisioned Ojai as the "glass community" he remembered from New England, but without the industry. On April 24, 1914, Libbey held a town meeting with 78 residents to discuss his plans. He hired architect Richard Requa to design a new town square in a Spanish motif, with mission-style arcades. He paved streets and renovated buildings. He sponsored a rat hunt by the local citizens to reduce the rodent population, offering prizes. The citizens cooperated in rebuilding the town, making their own investment. Libbey is believed to have spent over a million dollars in Ojai. He was not alone, as other industrialists such as Henry Ford, Thomas Edison, and Charles Goodyear helped out. Eventually Libbey added a country club, golf course, park, and library. By 1919, Ojai was a Palm Springs for Easterners. Libbey spent increasingly more time there each year until his death.

Libbey had an extensive personal collection of paintings in his home as well as in the museum. Like fellow Gilded Age collectors Henry Clay Frick and Andrew Mellon, Libbey loved to spend time with his paintings. The Dutch masters of the seventeenth century were popular with these collectors. Libbey had many of these masters in his West End home, which on his death would go to the museum. Some, however, he sold to other collectors. Meindert Hobbema was a favorite painter, and in 1914, Libbey added *A Pond in the Forest* to his collection. It was later sold and today is in the Allen Memorial Art Museum at Oberlin College. He liked such landscapes for the walls of his home.

Relaxation was a problem, with the ongoing development of new glass processes. Libbey's brother-in-law was probably his closest personal friend during the period. The Libbeys often vacationed with the Walbridges on the East and West coasts. Florence Libbey had her own set of personal friends with ties to the Scott family.

Libbey's work at the Toledo Museum of Art continued at a fast pace as well. The museum set another attendance record of 118,000 in 1914. It added an Etruscan and Greek pottery collection, which remains popular today. This collection of red-figured and black-figured ware illustrates the technology of early Greek pottery, and the early use of glazes and annealing. The museum's art committee — Libbey, Secor, and Spitzer — bought the collection to further explore the blend of art and technology.

Another industrialist, Alexander Humphreys, was fascinated by the interplay of art and industry. Humphreys was president of the American Society of Mechanical Engineers and the National Society for the Promotion of Industrial Education. In late 1914, he set up a truly massive exhibition at the Toledo Museum, one of many

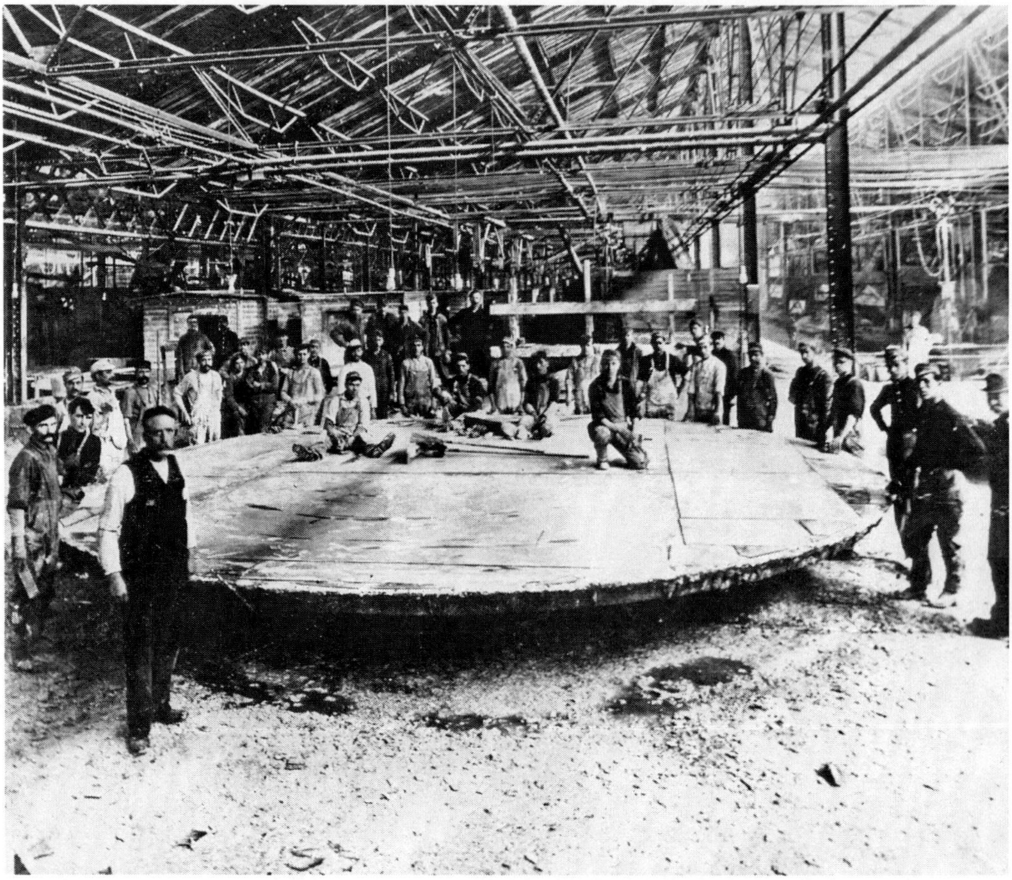

Grinding plate glass at the Ford glass plant (courtesy of Rossford Public Library).

people beyond Toledo involved in the museum. Alfred Duane Pell of New York donated a substantial rare porcelain collection to the museum in 1914. Eminent explorer and archaeologist Edgar Banks contributed the famous cylinder of Nebuchadnezzar to the museum. Even J.P. Morgan donated to the Toledo Museum. Libbey added what was considered the most important Hans Holbein work in America—*Portrait of Catherine Howard.*

Libbey's biggest project of 1915 was to establish an endowment and building fund for the Toledo Museum. The museum had operating funds for 1915, but there was no money for purchases and expansion. In the fall of 1915, Libbey offered to contribute $400,000 to an endowment if the citizens of Toledo would match it with $200,000. Treasurer I.E. Knisely, with his banking connections, immediately initiated a campaign. In a few days, without publicity, Knisely raised the $200,000. At the trustee dinner, a grateful Libbey announced: "In view of this public support and approval, without which no institution or movement can hope to exist or have success,

I wish to announce that I am prepared to give, in addition, to my portion of the endowment fund, the $300,000 or $400,000 necessary for the completion of the building."[7] Some of the funds were used to improve the central section. The endowment fund committee consisted of Libbey, I.E. Knisely, Jay A. Secor, Clarence Brown, and William Hardee.

Chapter 16

Working on a Legacy

In the years leading up to World War I, Libbey's life was changing. There was still much to do with the Colburn machine, but his wife was demanding he gear down a bit. He was trying to decouple from the intensity of business affairs but generally failing at it. It was a true struggle for him. Libbey loved new projects and the thrill of success, and in 1917, he even seemed to find new energy for business. Florence Libbey, however, hoped he would retire, allowing for more travel and vacation time in California and New York, and more personal collecting. Although the start of the war in 1917 had temporarily cut off European travel, Libbey spent a lot less time in Toledo, dividing more time between Ojai and New York. After the war, he spent less than four weeks a year in Toledo as he and his wife increased their European travel. While they maintained the Toledo home, the staff was released and they rented a suite of rooms at the Secor Hotel for their convenience. Often Libbey returned to Toledo alone for business. In 1917, he resigned as president of Owens Bottle to take a more active role in the new West Virginia-based flat glass company, Libbey-Owens Sheet. Libbey backed Clarence Brown to replace him, beginning a rift between Libbey and Owens.

The developing of the West Virginia plant of Libbey-Owens Sheet proved to be a nightmare. Libbey, unable to make his summer trip to Europe because of the war, became more involved in opening the new plant. The first problem was the loss of Irving Colburn. Earlier in 1917, Colburn was diagnosed with several health problems. Libbey made arrangements for him to take a vacation in Honolulu. The travel, however, did not help. On September 4, 1917, Colburn died. Problems at the West Virginia plant continued to mount, drawing Libbey into more involvement with operational issues. While Florence Libbey spent her time in California, Libbey was bogged down at Toledo Glass, Libbey Glass, Owens Bottle, and Libbey-Owens Sheet Company. While Libbey's attention and energy was drawn to Libbey-Owens Sheet, Libbey Glass was still giving him profitable and consistent returns.

The period of 1916 to 1918 was one of major breakthroughs for Libbey Glass and its development arm, Westlake Machine Company. Westlake had been working for years on development of an automated bulb-making machine, which it achieved in 1916, with a machine that could produce 10,000 bulbs per day. This rate was about twenty-four times the production rate of hand blown bulbs. The machine required close supervision and that added labor costs, but still it was more productive than the competing semi-automatic processes.

Libbey Glass had a semi-automatic bulb plant in Sandusky, Ohio, which it converted to a demonstration plant. As with the bottle machine, the company sold licenses. The licensee would pay a rental cost based on construction costs, and royalties from $1 to $2 per thousand "merchantable" bulbs produced. It was a significant royalty, considering the savings per 1,000 bulbs was around $5. General Electric became a licensee in 1918, buying the bulb business and the Sandusky plant. Libbey Glass was then out of the bulb business, but had profited handsomely from the development of the bulb machine. After four years, G.E. purchased the machines free of royalties for $450,000. Overall, it was another triumph for Libbey and for his selection of Solon O. Richardson to oversee the project. Richardson always maintained that he had a bigger role in glassmaking automation than he was credited for. Libbey had, in fact, overruled Richardson in the manner of selling the rights to bulb production, and it created another split among Libbey's companies. Libbey made the right decision, knowing that General Electric was pouring capital into Corning's development of a ribbon machine that could do 1,000 bulbs a minute. Libbey didn't have the money to compete with developing the ribbon process. Instead he needed capital to improve his bottle machine, to build a state-of-the-art bottle plant, and to engineer a new window technology plant.

Libbey and Richardson were not on the best of terms. Libbey had never forgotten how Richardson and Robinson opposed him in 1894. Libbey had always considered Libbey Glass as a producer of tableware and cut glass and did not want Libbey Glass doing development work. Libbey had arbitrated an awkward compromise back in 1908 with the formation of Westlake Machine Company. The owners of Westlake were Libbey, Walbridge, Donovan, Richardson, and Robinson. The first three were officers, leaving Richardson and Robinson with little role in the decision-making process. The legal agreement written up by Brown and Geddes allowed the transfer of patent rights for the Westlake machine to Owens Bottle Company, effectively taking Libbey Glass out of the bulb market. The Owens Bottle Company went on to block Libbey Glass from using the Westlake machine to produce tumblers unless royalties were paid. Richardson argued (correctly) that Libbey Glass had paid part of the development costs and got nothing in return. Libbey was president of both companies but favored Owens Bottle, and his closest associates — Brown and Geddes — stood with him.

Things got worse, as Westlake Machine had also developed the Danner process for tube making. This process was a breakthrough invention by Libbey Glass employee

Edward Danner, which cut labor costs on tube manufacture by 85 percent. Owens Bottle claimed ownership on the basis of the 1908 agreement, which said all rights on machines that could produce bottles would remain with Owens Bottle. This was a stretch, but Owens Bottle made the claim. Richardson would file a lawsuit against Owens Bottle in 1920. Libbey, while president of both companies, took to the sidelines, letting Brown and Geddes argue the legalities.

At the time, Libbey was looking to expand in and dominate the bottle market, so he did not want to overrule his Owens Bottle Company board. Clearly, Libbey let this go too far. As novel as the Danner process was, it supplied a bit of a niche market. Libbey was looking to consolidate in the bottle industry and enter into the window and plate market. He had been brilliant at tactical compromises between the egos of Owens and Richardson and the boards of two companies, but now he had to, in effect, take sides (while still appearing to be neutral). For Libbey, it was pure business. To Richardson's disappointment, Libbey remained neutral, allowing the disagreement to go to court.

The Libbey-Owens Sheet Company began construction of its new plant in Kanawha City, four miles east of Charleston in 1917. The land was between the Kanawha River and the Chesapeake & Ohio Railroad. Toledo's Devore–McGormley engineering firm was hired to design and construct the $1 million plant. The plant would have six Colburn machines to target various segments of the flat glass market. The Kanawha City plant would be state-of-the art and one of the most automated plants in the world. "Raw materials and finished glass are untouched by hand from the time material is received from the railroad up to the point of finished glass delivered at the end of the lehr ready for cutters."[1] The high degree of automation was directly attributable to Mike Owens, who had excelled at new materials handling at Toledo Glass. The real struggle at Kanawha City was getting construction materials during the war.

The opening of the plant in late 1917 brought new challenges, but Libbey's team was up to the challenges, and Owens was experienced in plant start-ups. Fowle had assumed a sales position and Bill Jacobs took the role of plant manager. Jacobs was a friend of Owens and, in retrospect, was a poor choice. He was a bottle man who struggled with flat glass operations, although he had worked on the Colburn machine development team. Jacobs apparently had Owens advise him on his contract, an eight-year contract for $5,000 a year with bonuses. At the end of the contract, he was to be given $200,000 in stock. Mike Owens was functioning as general manager of the operation, but he was also heavily involved with bottle machine development work back at Toledo Glass. Owens was not good at dividing his time, and that was becoming a problem. He continued to argue with Libbey about the involvement of "damn lawyers." Bill Jacobs had Owens's support throughout the difficulties. Arthur Fowle took heat for his construction failures. Libbey and the Glass Board grew impatient and frustrated as problems continued.

Customers began to pressure Libbey and the board for a change. The board could not deal with Owens, and Libbey was forced to get involved personally. Libbey was always the best at deciding when to use the lawyers and when to use his operations men. Libbey also understood the weaknesses of Owens better than Owens himself. Clarence Brown pushed Libbey on the need for change, and Libbey considered Brown his most trusted associate. Brown had the ability to tell Libbey when he had made some bad decisions. In December of 1917, Libbey toured the plant and decided a change was needed. He was clearly dissatisfied with the overall performance of Owens, Jacobs, and Fowle. Libbey suggested that Jacobs be replaced with an experienced flat glass manager. Libbey and Owens agreed that Owens would do best back at Toledo Glass, where some improvements to the bottle machine were underway. Owens resigned as the general manager of the plant and Jacobs was fired.

Libbey named an experienced flat glass manager, Seth B. Henshaw, general manager of Kanawha. He had been recommended to Libbey by a number of customers, but he had a Mike Owens personality. Owens was jealous of Henshaw, but would soon have an ally in James Blair — son of Alfred Blair — who didn't care for Henshaw. Henshaw was hardheaded and obstinate, but, like Owens, he was considered brilliant. Henshaw proved to be the turnaround manager needed for the job. Henshaw again was a pick that few men except Libbey might make, but he was as successful as Owens, Richardson, Stevens, and the other non-conforming managers that Libbey chose. Henshaw was the perfect fit for the world's largest and highest quality flat glass factory. He instituted up-to-date quality control methods, and increased wages and incentive programs, which led to profitability.

Owens was put on development work at the plant, using machine #6, and worked on a number of new products such as optical glass. Libbey assured Owens that he had total control over machine #6. Owens and Henshaw clashed, but the plant turned profitable in 1918. Owens had total control of machine #6 so that he would not have to deal with Henshaw, avoiding any sparks. The board directed Owens to spend all of his time and effort on machine #6, virtually imprisoning him there, away from Owens Bottle and Toledo Glass. Unsurprisingly, after he turned the plant around, Henshaw was fired because of problems with his personality. He did finish out his initial three-year contract of $10,000 a year.

On July 30, 1918, Clarence Brown died. It was a personal blow to Libbey. Brown had been his closest associate, who served as a director at all of Libbey's companies with the exception of Libbey Glass. Brown was active on a number of museum committees. He was a skillful executive able to cope with many large egos, family ties, and difficult people. Behind the scenes, Brown and Libbey owned a great deal of Toledo real estate. No man was closer to Libbey on so many levels. Through his many investments, Brown was one of Toledo's richest citizens, as well as a great leader and philanthropist. His millions would be tied up in lawsuits for years.

The man who replaced Brown in his directors' chairs was James Blair, whom Brown had brought into Owens Bottle. Blair was the son of Toledo industrialist James C. Blair, who had been president of Wheeling & Lake Erie Railroad. Blair graduated from the University of Michigan in dentistry and was often referred to by Mike Owens as the "damn dentist," although Owens came to accept Blair. In 1916, Libbey had designated Blair as Owens's "understudy." Libbey had a great deal of confidence in Blair who was one of the few people with whom Owens avoided an argument. Blair's temper was more firey than Owens's, and Blair could focus it with force. More importantly, Blair proved to be an outstanding executive.

In 1918, Blair was vice president of Libbey-Owens and Owens Bottle Company. At Owens Bottle, his lieutenant was John Biggers, and at Libbey-Owens, it was William Boshart. Both Blair and Owens were vice presidents, and there was a love-hate relationship between the two. Blair also carried the title of general manager at both companies, which Owens did not. Libbey, as was his style, tried to tacitly manage the two for the good of the companies. Libbey "informed Owens that he would enjoy free access to Libbey-Owens' factory but that he should never give any orders there directly to anyone and should discuss all matters with Blair. Libbey then suggested to Blair that he always consult Owens on matters pertaining to company policy."[2] Libbey was able to cope with the organizational stress to achieve his goals. Owens fought hard with the board of Owens-Bottle to regain full authority. He started to complain that Libbey's winters in California, summer trips to Europe, and the expansion of Libbey-Owens Sheet left Owens Bottle without a decision-maker. In actuality, Libbey and Brown and then Blair kept in close contact via telephone and telegraph with the operations. Owens was feeling more and more isolated.

Libbey was now deeply involved on many business fronts. He was building a bottle monopoly, entering the flat glass segment, and automating bulb and tube production at Libbey Glass. Sometime in 1912, Libbey went after domination of the bottle-making segment of the glass industry. Owens Bottle had about a 50 percent share of the market. As Mike Owens had predicted, the bottle industry was booming as glass packaging penetrated new products such as beer and soft drinks. Libbey and the Owens Bottle Board put together an aggressive plan to control the bottle market, as Libbey felt bottle production was necessary to maintain a future income for Owens Bottle. The CX machine development and the formation of Libbey-Owens Sheet Glass had slowed Libbey's bottle plan. Still, in 1916, Libbey took over American Bottle Company, which doubled his bottle production. American Bottle was one of the first licensees of the Owens bottle machine. Its forty-six machines at plants in Streator, Illinois, and Newark, Ohio, produced beer bottles for most of the Midwest. American had a capacity of 324 million bottles per year. Owens Bottle took over complete stock control but allowed the company to operate under the name American. Libbey didn't stop there but continued his greatest effort to create a horizontal monopoly. In

1917, Owens Bottle took control of Graham Glass, which was the major bottle producer for soft drinks such as Coca-Cola.

This was one of the rare times that his associates viewed Libbey's plans as grandiose and Napoleonic. Libbey showed this side in a letter to Clarence Brown when he dreamed of controlling Illinois Glass (a major whiskey ware producer), Hazel-Atlas (home canning jars), and Ball Brothers (fruit jars). Libbey noted the advantages:

> Our most favorable channel through which to expand is, in my opinion, the acquisition of the other large U.S. companies — Ball Brothers, Hazel, and Illinois ... if secured upon a fair basis ... would give our company not only supreme control of all branches of the bottle trade, but to its stockholders, a total of $6 to $7 million earnings, which if proven true would exceed my previous year's estimate of $5,000,000 net profit in all branches of the bottle business under the controlling influence of the great and wonderful invention of ours. I hope to see this a fact soon and I shall then feel that it was worth living for, to be one part of the mechanism that gives its stockholders, its original inventors, and promoters a keen satisfaction in its success physically and financially and greater than all to be instrumental in giving to the world cheaper bottles when the time for consolidation may come.[3]

This seems like a letter he would have written twenty years earlier.

Libbey's plan certainly might run into problems with the Sherman Anti-Trust Act, but he wanted to move forward. But the consolidation never took place. Libbey next considered a bottle plant in Charleston, West Virginia, in 1917. This bottle plant would have six machines and automated handling systems. Libbey's campaign to corner the bottle market might explain his decision to take Libbey Glass out of bulb production. The Westlake machine rights sale provided an infusion of capital for both Owens Bottle and his new flat glass company.

Owens Bottle Company was also aggressively expanding its license agreements, especially with its new AV machine. The AV machine was a 15-arm monster that could produce 625 gross of pint beer bottles instead of the 500 produced by the 1914 model. It doubled the production of earlier machines. But after the development work was completed Mike Owens was again at loose ends. He was functioning as vice president of Owens Bottle and was handling sales expansion. Owens would continue to be a thorn in Libbey's side when it came to customers and sales. Owens almost lost a major licensee in Ball Brothers. There were five Ball brothers. The youngest one was Frank, who was president of the Muncie, Indiana, company. Frank Ball was a lot like Mike Owens and that soon became a problem. Frank was hard-driving, persistent, and hardheaded. He had demanded trial production of various size jars, requiring set production rates. As Frank Ball suggested more trials, Owens's patience wore thin. Owens reportedly told his associate, "Tell Mr. Ball never to come back here again. We are through with him." Eventually cooler heads prevailed, but this was typical of Owens. Owens didn't believe in dealing with problem customers.

He saw them as costly and time-consuming. Libbey agreed to a point but was much more diplomatic and pragmatic. Libbey often had Walbridge run interference between Owens and customers.

Libbey had more problems with Owens in late 1917. Owens was needed for developmental work at Toledo Glass and at Owens-Libbey. In addition, Owens was still a key officer at Owens Bottle, and some developmental work was under way there too. Owens Bottle, under Owens, was expanding bottle production. For several years, Libbey had been concerned about the competing Graham bottle machines and others, and he wanted Owens to focus more time on the bottle machine improvement. The Owens Bottle Company board wanted to reduce Owens's corporate role as well. Libbey was convinced that he had to keep Owens out of direct management of the various companies, if possible. Owens served Libbey best on development project work. Owens Bottle Company was heavily involved in production, sales, and market expansion. These areas were not Owens's strength. Libbey and the board started a search for an "understudy" to Mike Owens. The short-term goal was to free Owens for development, but clearly they saw the person as a long-term replacement as vice president for the temperamental Owens. For his part, Libbey was moving on, no longer wanting to be involved in the day-to-day battles into which Owens dragged him into at times.

Libbey's relationship with Owens was complex. Many called it a father-son relationship, but that model would not explain it fully. It is better explained by Libbey's extremely deep sense of loyalty. Libbey was loyal to those who had been part of his success, like Owens. Libbey also held those in high regard who demonstrated loyalty to him, which Owens always did. Loyalty appears to have been the basis of their relationship. Libbey and Owens were not social friends. Owens always showed Libbey respect. To Libbey, Owens always represented earning potential. The Owens Bottle Company carried life insurance on Owens.

The process of hiring an understudy proved difficult, as Owens and the board disagreed on the type of person needed. Libbey had earlier tried to make James Blair the understudy, but Libbey feared losing Blair because of Owens. They were interviewing young engineers and managers from big-name colleges. Owens would have none of it. Finally, after turning down a candidate from MIT, Owens agreed on the young assistant treasurer of the company, John (Jack) Biggers. Libbey approached Biggers with the following offer:

> Jack, Mike has paid you a great compliment, but it is a dangerous one. You know I've been trying for years to get him to select an assistant as an understudy. I picked out fine men and recommended them to him. He said, "Mr. Libbey if I've got to have an assistant I'll take Jack Biggers," that was a surprise to me, and it may be to you, but I want you to think it over very carefully. As I said, it's a compliment but it could be the reverse because you know how arbitrary he can be especially with subordinates, and if you dis-

> pleased him, and he fired you, I could never take you back with the company. You'd be through. So, Jack, you think this over very carefully. Talk it over with your family.[4]

Biggers went on to accept the position as assistant general manager, where he experienced many rocky times, but earned the respect of both Owens and Libbey. Biggers was able to grow at Owens Bottle when in 1918 Owens was sentenced to development work in West Virginia.

The successful year of 1918 ended with an agreement with a Japanese company for exclusive rights to the Colburn machine in Japan. Gosaburo Sugita of Japan-American Glass had come to Libbey to acquire rights for the Owens bottle machine, but was also impressed with the Colburn machine. Libbey was able to sell first rights to Sugita. Libbey-Owens became a stockholder in the Japanese company. As a reward, Libbey sent Otto Miller, Irving Busard, Arthur Fisher, and Clyde Belden and their wives to install the machines in Japan. All of these were part of the original development team. The 1919 Libbey-Owens stockholder meeting at the Nicholas Building was the largest to date. With the war over, flat glass was in high demand, and no one had anything remotely close to the productivity and quality of the Colburn machine. An even brighter horizon awaited Libbey-Owens as a supplier to the automobile industry. One industry journal noted:

> At the present time, the Libbey-Owens Sheet Glass Company can fill any orders for any quality, and any size and thickness, that any other window glass factory can fill, and besides their product has the preference of the trade, for the following advantages over cylinder-made glass: it is perfectly flat, uniform in thickness, the most perfectly annealed glass made, due to its method of manufacture, lack of distortion common to all cylinder-made glass and is much in favor by automobile manufacturers for windshields and closed cars, for that reason.

Henry Ford was demanding a new type of safety glass for windshields, and Libbey had just the man in Mike Owens to tackle such a development program. The excitement of the project seemed to once again energize Libbey and his old warhorse Owens.

In 1919, Libbey was still president of Libbey Glass, Libbey-Owens Sheet, and Owens Bottle Company, functioning as what we would call a CEO today. He made operating decisions on the bigger projects in all the companies but was trying to back out of day-to-day operations. Solon Richardson was operating head of Libbey Glass. James Blair and Owens were vice president of Libbey-Owens and Owens Bottle respectively, but James Blair was in operating control at both companies. At Owens Bottle, Vice President John Biggers, who also functioned as assistant general manager, backed up Blair. Officers and directors included seasoned managers in Boshart, Fowle, Geddes, Schmettau, and Walbridge. Libbey had done what many industrialists failed to do — develop an organization of extremely talented executives and managers.

Libbey's passion for art and glass collecting increased, even though the war restricted travel to Europe. He proved to be more than just an art collector. From the early beginnings of the museum, he encouraged artists. Each year the museum had a local artist exhibition and a local photographic exhibition. While many collectors such as Henry Clay Frick and J.P. Morgan focused mainly on European painters, Libbey had an interest in new American painters as well. In 1909, Libbey was a supporter of George Bellows, who at the time had no national following. He bought *Blackwell's Bridge*. Now in 1916, Bellows had a national following. By the 1920s, George Bellows was in all the major American museums. Libbey did the same for Ralph Blakelock. When Libbey purchased his first Blakelock, the artist was impoverished. In 1916, Libbey purchased Blakelock's *Brook by Moonlight* at a New York auction. At the time, the $20,000 price was the highest ever for an American painting. The sale was negotiated through Reinhardt galleries. By 1916, Blakelock had been committed to an asylum. Libbey's purchase focused attention on Blakelock. Libbey loaned his canvas to several museums and created a fund to support the artist in his last years.

Libbey was anxious to expand his glass collection and added a 450-piece collection from eminent Philadelphia authority and collector Edwin A. Barber in 1918. The Barber collection included some early American glass by Baron Stiegel and Casper Wistar, the finest examples of American glass from the 1700s. The collection also included some glass pieces from Jamestown in the 1600s. With these additions, Libbey's glass collection represented the integration of art and manufacturing. The collection also included examples of Millefiori glass, Italian for a "thousand flowers," as well as flasks belonging to Lafayette, Franklin, and Washington. Libbey now had one of the world's most representative glass collections, and it was a matter of great pride for him.

Chapter 17

The Taming of the Lion

With the war over, Libbey was able to get his West Virginia flat glass operation running at capacity. Libbey's decision to hire Seth Henshaw had been successful, in spite of his personality problems. All six sheet machines were running quality product, and Owens was making developmental progress on automotive glass, which loomed as a giant new market. The stock of Libbey-Owens Glass had rocketed from $30 a share to $180 a share. Libbey-Owens Sheet Glass reported an astonishing profit of over $4 million. Other things had changed. With Brown gone, Libbey was dealing with second-generation executives. Even Owens was now talking about letting the young guys run the company, but neither Owens nor Libbey could pull completely away. Owens, however, could no longer count on calling Libbey to make some mid-level decision. At this point, Libbey wanted the freedom that his chain of command gave him.

Libbey now hoped to sell the Colburn process to Europe. With the West Virginia flat glass plant under control, Libbey planned a major business trip to Europe. He was to take Charles Schmettau, vice president of Libbey-Owens. Schmettau was an experienced contract lawyer who had worked in the law firm of Brown and Geddes, and had demonstrated sales ability as vice president. Libbey would need both skills in this new endeavor. Schmettau was also the museum's secretary, a trustee, and a member of the museum's executive committee, making him the ideal companion for a European trip. Libbey and Schmettau left in the summer of 1920 for Europe, where money was tight as the countries were struggling to pay off war debt. European wages for decades had been about 60 percent less than those paid to American skilled workers, which made automation less attractive. It became clear that they would need to put together a consortium of companies to finance the new flat glass production. Schmettau carried the load, as Libbey often went off art collecting. Of course, Schmettau did a bit of collecting himself. Libbey and Schmettau put together a joint venture in Belgium, France, and England, which again was a brilliant business success. They started to build a plant with 12 Colburn machines in Belgium, with the idea of expanding from there. Libbey-Owens Sheet Company got a combination of stock and royalties. Similar deals were set with Canada and China.

Libbey's personal profits boomed in the later part of 1920. His companies across the board had plenty of cash for expansion. Libbey-Owens declared a 25 percent stock dividend for board members and stockholders and a 10 percent cash bonus for staff employees. In addition, Libbey announced a long-term stock plan for employees, institutionalizing a system that had brought him so much success. The original Toledo entrepreneurs had prospered amazingly from 1888 to 1920. Scoville noted their position in 1920: "The same entrepreneurs who by investing $175,000 launched the Libbey Glass Company and Toledo Glass Company in 1888–1895 still retained controlling interest when the combined assets of their enterprises amounted to between $50,000,000 and $60,000,000."[1] Of course, Libbey, Owens, Brown, and Geddes, and to a lesser degree, Richardson and Robinson, had prospered greatly. In 1920, Libbey finally allowed his old friends Richardson and Robinson to invest a small amount in Libbey-Owens. The second generation, like Blair, Schmettau, and Biggers, also could envision amazing fortunes from their investments. Libbey again proved brilliant at grooming this second generation of executives. No other group of business partners had profited as much, with the possible exception of the members of Andrew Carnegie's steel group. Most of Libbey's early partners were now living in Toledo's West End, which was full of mansions, art collections, and new cars.

After the war, Libbey started to slow down a bit, with a new focus on developing Ojai and the Toledo Museum, which had 114,000 visitors in 1921 and offered over 200 lectures. The period after the war was one of great excitement. The museum's School of Design was opened in the old home of William Scott, Florence Libbey's uncle. The museum was on sound financial ground with interest from its endowment fund paying half the operating cost, dues paying another quarter, and the city paying a quarter. Libbey devoted more time to his personal interests, although he continued to head the boards of his companies, and remained president of the museum until his death. He made more trips to Europe for fun, art collecting, and business. He remained at the head of Libbey-Owens Glass, guiding it into a new era of glassmaking. Libbey's schedule, while including many vacations, would have been difficult for a man half his age. He proved a master at mixing pleasure, work, and hobbies.

The king of Belgium visited the Toledo Museum of Art on Oct. 7, 1919. Belgium had suffered much during the war and aid poured in from many quarters in the United States, including Toledo. George Stevens was, like most of Toledo's residents, deeply involved in the war effort. The museum's William Scott house was a Red Cross headquarters. Libbey and thousands of Toledo children donated to the Belgian Orphan Relief Committee. A former mayor of Toledo, Brand Whitlock, was ambassador to Belgium during the war. The king presented the museum one of the sketches for *Saint Martin Sharing His Mantle with the Beggar* by Sir Anthony Van Dyck. It was a particularly fitting gift to honor the aid given by Toledo residents. The sketch had been in a vault for four years, awaiting safe passage for King Albert.

Libbey made a major addition to his glass collection in 1919 by purchasing nearly 5,000 pieces from New York City banker Thomas E.A. Curtis. It was considered the largest private collection of antique glass in the world. On Curtis's death, it was to be auctioned at Anderson Galleries in New York. The auction would have scattered the collection around the world, so Libbey moved quickly to purchase it directly from the family. The Curtis addition nearly doubled Libbey's collection and cost him $121,225. The acquisition caused headlines, as Curtis bought most of the pieces through an agent, Azeez Khayat, who had dealings with grave robbers in Syria, Turkey, and Palestine. Such purchases would be banned today under international law. These pieces were extremely attractive to Libbey because they were examples of unique periods and processes. For example, there were fourth century sand cast (sand core) vessels from Syria and hand-blown art pieces from the Middle East. The collection had some of the earliest examples of Millefiori (thousand flowers) and ribbon-colored glass as well as fused inlay mosaic. Libbey would make the collection a gift to the museum in 1923. Stevens noted: "Henceforth all students, writers and connoisseurs interested in glass in the arts, will make Toledo their Mecca."[2]

After the war, Libbey returned to making summer trips to Europe. Between Europe, Ojai, and Florida, the Libbeys were in Toledo less than four weeks a year. When in town, they rented a suite at the Secor Hotel, as did many old West Enders. Often Libbey returned alone while Florence remained in California or Europe, where the Libbeys had many old friends. Carl Spitzer and his wife had retired to California and also maintained a suite of rooms at the Secor Hotel. Libbey would join Spitzer in Europe for auto trips and art collecting. European trips always included Paris and Switzerland, where Florence Libbey often stayed in Geneva with her friend, Nettie Poe Ketcham, who was living there. Each year the Libbeys would meet the Stevenses and other Toledo neighbors. George and Nina Stevens spent most of the summer (the museum closed in summer) at an art colony in the Netherlands and then would meet the Libbeys in Paris for art purchases.

Art collecting and the development of Ojai became Libbey's primary outside interests.

The museum board developed more community diversity as the museum grew into a new phase. The 1920 museum trustees included Jay Secor, Bill Gosline, Isaac Knisely, Arthur Secor, Edward Ford, Emil Bock, William Hardee, the Rev. George Johnson (head of Toledo's catholic schools), and Allen Stockdale. On January 9, 1920, Florence Libbey and the trustees had a recognition dinner for fifty of Libbey's closet friends. Out-of-town guests included Henry Reinhardt, John Beatty (Carnegie Institute), Henry Bailey (Cleveland School of Art), Ralph Booth (Detroit Institute of Art), and old friend Frank Gunsaulus. Libbey was at a point when most people look back on their lives, but that was inconsistent with his character. He was planning new challenges.

Another important event of 1920 was the unveiling of the John Burroughs sculpture at the museum. C.S. Pietro, who had come to the United States originally to do a bust of J.P. Morgan, did the sculpture. The bronze, life-size sculpture was mounted on 20 tons of granite. John Burroughs was a great naturalist whom Emil Bock had admired for years. Bock commissioned the sculpture for the museum. He had left Toledo Glass years before but remained a friend of the museum and Libbey, and was a very successful inventor and industrialist in his own right. Both Pietro and the aging John Burroughs were at the ceremony, which included 15,000 school children. The statue was consistent with the evolving role of the museum and the beautification of the city.

Libbey was becoming highly revered in art and museum circles for the creation of a model community museum. He was asked to speak at the Metropolitan Museum in New York on the relationship between the museum and art education. In fall 1921, the Congrès de l'Histoire de l'Art in Paris asked him to speak on the topic. Libbey was proud of the unique alliances and approaches the Toledo Museum had forged. He summarized his view in Paris, that the museum intended "to bring to our citizens the understanding of the principles and the benefits of art in their lives and in their work. To this end our mission of art has within the past few years become active educational institutions, as well as safe and necessary repositories of art." Libbey continued, "The burden has developed upon private citizens to create, endow, and maintain our institutions of art until such a time — and we hope that time will be soon — as our government will recognize the value of art education in the lives of all our people that may be able to observe, to understand and to create beauty."[3] A few months later, the city of Toledo gave the museum a $15,000 appropriation. This trip was one of Libbey's proudest moments. He had created a true art-industry-community alliance that was seen nowhere else. Libbey continued to expand the museum's educational ties in the 1920s, setting up a partnership with the University of Toledo to offer college credit courses.

The museum continued to expand its art collection, with Arthur Secor and Carl Spitzer making major contributions. Secor, in particular, added a huge and valuable collection in 1922. Some of the collection had hung in his West Side home for years. This donation included two paintings of Alexander Wyant and George Inness. Other notable works included those of Diaz, Blakelock, Millet, Corot, Mauve, Israels, Rousseau, and Maria. The Toledo Museum was selected for several touring international exhibitions.

Stevens developed the School of Design, with his wife becoming the first director. Stevens and Libbey had always envisioned the museum leading a revolution in design and industrial design. The museum, from its inception, was a tribute to manufacturing in all the arts, with classes in photography, pottery-making, needlework, drawing, painting, lettering, and home decoration, but now Libbey wanted to define the link between art and industry. The design school was located in a large mansion, the former home of Florence Libbey's uncle, William Scott. Mrs. Ellen Meeham, a graduate of the Boston School of Practical Art, was named supervisor of the design classes. Miss Maud Streeter, a graduate of Pratt Institute, was hired to manage applied design classes. The school was equipped with kilns for pottery-making, looms, potters' wheels, drafting equipment, and art supplies.

Classes were free to pupils of the Toledo public and Catholic schools. However, the focus was also on workers, engineers, and industrial designers. It was to be a new alliance of art and industry, the beginning of a movement that would grow in the 1930s. The concept of an industrial designer had evolved under Libbey at Libbey Glass. When the company was being run by his father, Libbey had organized pattern design contests; this school would spread that his concept to the public. The School of Design had the enthusiastic support of the new assistant curator, Blake-More Godwin. Godwin would years later note: "the effectiveness of museums lay more in their impact on manufactured products than in the acquisition and preservation of works of art."[4] This reflected Libbey's vision of art and industry, going back to the design prizes offered to employees in Boston.

The long feud between Richardson of Libbey Glass and Michael Owens continued. Court proceedings started in late 1920. The original agreement was that Owens Bottle owned all rights to any equipment developed at Libbey Glass that could be used for bottle-making. Libbey Glass had lost the battle over the bulb making process to Owens Bottle, and the battle continued. Libbey had clearly let this go too far, but now his interests and fortune were tied to Owens Bottle and Libbey-Owens. He had even reduced his stock position at Libbey Glass but had remained president. Richardson had been pushed too, as Libbey Glass had developed a tube-making machine that Owens Bottle claimed to own because it could be used for bottle making. Richardson was clearly right this time, but the argument was an old one between Owens and Richardson.

With some truth, Richardson claimed to be part of the development of the Owens bottle machine, but the feud really went back to the bulb-making days in Findlay in 1892. It had been bitter, with Richardson and Owens fighting for Libbey's favor like two rival siblings. Richardson had lost that favor in 1893 when Owens helped Libbey achieve the World Fair success over Richardson's objections to participation. That had cost Richardson future involvement in some of Libbey's better investments. Now, in 1921, Owens and Richardson were in separate corporate boxes and were not on speaking terms. The Owens Bottle board, linked to Libbey-Owens, was on the side of Mike Owens. What Libbey had allowed in order to retain two of his best people was now coming back to haunt him.

This final argument was over a glass tube machine developed by and paid for by Libbey Glass under Richardson's guidance. Glass tubing was one of the earliest products of Libbey Glass, going back to New England Glass. Richardson himself had sold glass tubing to Thomas Edison back in the 1880s. Prior to the Danner machine, tubing was hand pulled or stretched out by men walking it for yards. The automatic machine could produce 2,600 pounds of tubing in one day, or the product of ninety men working by hand for a single day! Libbey Glass had licensed its use by Kimble Glass of New Jersey. Owens Bottle argued that tubing could be used to make a bottle-type product; therefore, Owens Bottle owned the rights to the machine. It was a weak legal point that tubing could be used to make bottles, as alchemists had done in the seventeenth century. Libbey should have weighed in with Richardson and Libbey Glass; instead he took a neutral position.

Richardson stated in his deposition:

> We have visited Mr. Libbey, stated the facts to him, and asked him to use his good offices to straighten the matter out. His reply is that he is interested in both companies and must be neutral, when, in fact, he was President of both Companies and verbally made the agreements with the Libbey Glass Company's officers who are actively in charge. He does not deny the statements, but claims there are many new stockholders in the Owens Bottle Machine Company.[5]

The trial started before Libbey could get out of the country to avoid being involved. A witness reported Libbey's awkwardness:

> The court had recessed for the noon meal. Mr. Owens and the witness were the first parties concerned in the trial to reach the dining room of the Hotel Secor where they were at a table together conversing when a group from the Libbey Glass Company came in and sat down at another table. Nods only were exchanged. The New Jersey group came in and sat down at another table, then some men from the Owens Bottle Machine Company likewise. Finally Mr. Libbey arrived for his meal and paused at the door of the dining room to choose a table. His eyes swept the room with its several sets of diners; then he deliberately walked to a small table against the wall and sat down by himself with his back to the whole room. There he dined alone, and no man spoke to him or recognized his presence other than covert glances until Mike Owens impulsively arose, walked across the room, put his arm affectionately around Mr. Libbey's shoulders and bent over and conversed privately. The rest finished their meal and filed out without a greeting.[6]

In the end, Libbey Glass won the lawsuit.

Libbey-Owens Company was in flux between 1920 and 1922. The company was in a position to take[7] over and dominate the flat glass market. It had license agreements in Europe and nonexclusive agreements with a number of plants, including Pacific Libbey-Owens Sheet in Los Angeles. Pacific was a joint venture, with Libbey-Owens owning 51 percent. The company was highly profitable, and John Blair and Mike Owens proposed some changes that required Libbey's personal attention. Owens had been taken off his developmental work on machine #6 at the Kanawha City factory so the machine could be used to meet growing window glass demand. Seth Henshaw believed he needed it to meet new orders for window glass. Owens believed the future of flat glass was in the booming automobile industry. He had been working on a difficult problem with machine #6. Colburn glass had one weakness, glazing or distortion in slight waves on the surface of the glass. A moving car intensified the effect. The problem caused eye fatigue for the driver. Ford Motor and others had turned to plate glass manufacturers who used a grind and polish process on cast plate. This was the type of technical problem made for Mike Owens. The Libbey-Owens Board, however, was happy with the full capacity and high profit of window manufacture. But Henshaw had gone too far in his disagreements with Owens. Blair, who had battled Henshaw for months, would remove Henshaw and replace him with Thomas McKinley.

While they had separate boards, Libbey never had fully separated Libbey-Owens and Owens Bottle Company. The two companies shared executives and many functions. For example, the purchasing, accounting, and traffic departments of Owens Bottle performed the same functions for Libbey-Owens. This cooperation kept costs and overhead down. It also gave the two companies more leverage in large purchases and shared shipping. Libbey had created many monopolistic advantages without the scrutiny of the trust department. As for his role in Libbey Glass, the company reorganized in 1920 with his old friend J.D. Robinson as president. Libbey Glass had been in the background since the 1890s, but from 1890 to 1920, it had increased its sales elevenfold, its earnings tenfold, and its assets increased by a factor of 26. All of this greatly contributed to Libbey's personal wealth.

Libbey and his limited stockholders in the various companies were extremely wealthy people. Libbey Glass, Owens Bottle, and Libbey-Owens were highly profitable. Still, an amazing amount of wealth also was centered in Toledo Glass, which owned many patents and collected royalties. The core stockholders of this Toledo Glass group were Edward Libbey, Clarence Brown, Frederick Geddes, W.S. Walbridge, and Mike Owens. These five men reaped the benefits of the development of automatic glassmaking. John Biggers and William Boshart joined the group later, but even they became very wealthy. Toledo Glass Company stock was not traded on the open market, but some sale values were recorded. One employee who worked on the bottle machine development was rewarded with 50 shares at $150 per share in 1900. Ten years later he sold his shares for $800 per share. In 1925, the shares were estimated for Libbey's estate at $9,000 per share. Glass historian Warren Scoville reports that at the end of the 1920s, the value was $14,000 per share (140 times the original value).[8] With the exception of Mike Owens, these men invested in many of the related companies, multiplying their fortunes. Owens, however, sold stock and spent his money on fine living, which concerned Clarence Brown.

In 1920, Libbey-Owens reported an amazing profit of over two million dollars, and was constrained only by the need for more machine capacity. The directors declared a 25 percent dividend on preferred stock. Once again, investors in Libbey's projects were highly rewarded. Libbey was now ready to make some final adjustments in the Libbey-Owens management. On July 12, 1921, Libbey-Owens took over United States Sheet & Window Glass Company and started plans for a new plant at Shreveport, Louisiana. A few weeks later, Blair fired Henshaw and replaced him with a veteran from outside the company, Thomas McKinley. Libbey-Owens also took over Fairfield Glass of Lancaster, Ohio, and installed a Colburn machine there. In addition, Libbey-Owens expanded the West Virginia plant, making it a major force in the flat glass market. By 1923, Libbey-Owens had six Colburn machines at the new Shreveport factory, 18 in West Virginia, and one in Lancaster. The company had the capacity to produce four million boxes of sheet glass a year, which was four times its capacity in 1917. (A box contains 100 square feet of glass.) The offices of Libbey-Owens were moved to the twelfth floor of the Nicholas Building on Madison Avenue. Libbey was pleased that he had moved from development to commercial domination of the flat glass industry.

Mike Owens was able to get a look at the grind and polish process at Edward Ford Plate Glass across the river at Maumee. A friendly relationship between Ford and Libbey that paved the way for Owens, even though the Colburn process was a direct threat to the grind and polish window business. Ford was a trustee and committee member at the Toledo Museum. Edward Ford was a co-founder of Pittsburgh Plate Glass and had started his own plate glass operation in 1899. Owens felt he could add a simple grind and polish to the Colburn process. He was overruled, but as in the past, appealed to Libbey. It was a struggle, but Libbey finally agreed to allow the project to be financed.

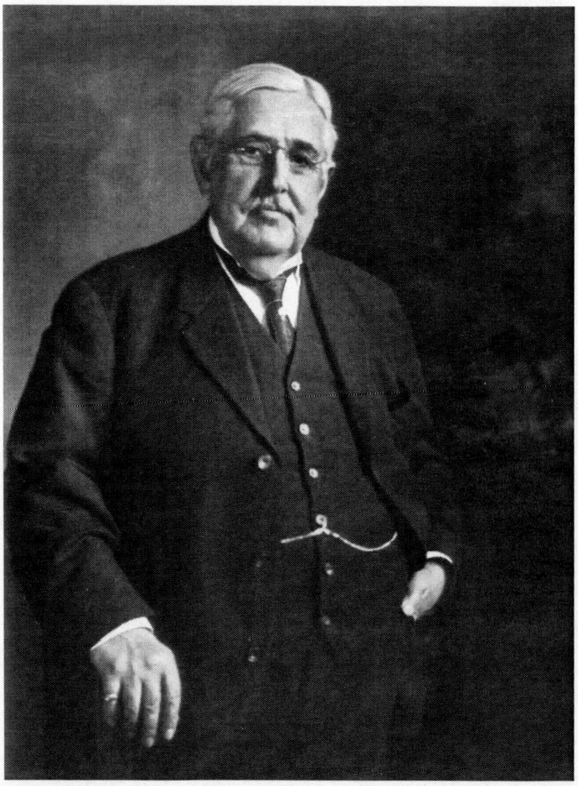

Edward Ford, 1910 (courtesy of Rossford Public Library).

Owens moved quickly to assemble a team and start a small experimental operation in a space adjoining the Kent-Owens Building. At work were Joe Crowley of the old Colburn machine team, Edgar Peters of the Owens Bottle engineering group, Bill Smith (Owens's cousin), and a group of master mechanics. They made quick progress thanks to the outstanding work of Joe Crowley. The Libbey-Owens board voted to expand the experimental plant for production. Owens continued to work on a number of other projects, including safety glass production and bottle machine improvements. Even at 64, Owens did not slow down. He had found happiness in his work and a peaceful life in Toledo's West End. He avoided high society. He played golf at Inverness and took winter golf trips to Florida. Usually the Libbeys would have Owens and his wife out to Ojai for a visit, as their son lived in Hollywood. Owens spent his evenings playing solitaire or reading from his extensive library. He enjoyed a Sunday drive and had become a bit of an amateur Civil War historian.

Owens was pleased that the Libbey-Owens Board approved the building of a Toledo factory to produce automotive glass in 1923. He had worked hard to develop this new cast-grind-polish system that had excited Henry Ford. It would allow him to spend more time at home. In 1922, Owens pushed the board, but with so many expansion projects underway, it lacked the capital for another $2.3 million-plus plant. Construction started in mid–1923. The plant capacity was to be 15 million square feet of plate glass.

While construction was underway, Owens initiated a new development project, to make safety glass. Again, Owens was looking to the future. James Blair and Libbey were behind this project and got Owens the funds to start development. Once the Toledo plant was completed in 1925, Libbey-Owens had 40 percent of the flat glass market. James Blair was a dentist with little glass experience, but Owens came to trust him. Blair also came to have deep respect for Owens. This new relationship took the pressure off Libbey to be the mediator between Owens and the board. One thing Libbey and Owens always agreed on is that it was the person, not the education or experience, which was important in the selection of executives. This strategy had failed them at the plant manager level, but for the board executives and the directors, it proved an excellent approach.

Chapter 18

Final Months

Libbey's impact on the glass industry by 1925 was astonishing. He had changed the way glass was made for 3,000 years. From 1900 to 1925, glass production had doubled due to Libbey's lower-cost, automated processes. Technology and new markets had caused a 39 percent increase in glass employment during the same period. America was producing 500 million square feet of window glass, 94 million square feet of plate glass, 288 million glass containers, and 70 million pieces of glass tableware per year. Even with all the automation, Libbey had achieved a peaceful transition in the glass industry without the labor upheaval of other industries. Glassworkers still were some of the highest-paid industrial workers in America. Politically, Libbey had proven that tariffs could increase employment without excessive price increases. He set the ideal for corporate responsibility in reinvesting profits into the local community—both with factories and civic projects. At Libbey Glass, he had managed to industrialize craftsmanship. He had changed the industry and the country. Libbey's passion in his final years was the Toledo Museum of Art and the development of Ojai, California, but he never fully gave up running his glass companies.

While Libbey was only spending a few weeks a year in Toledo, he still had a great love for the city, although he also was involved in the continuing development of Ojai. In January of 1923, Libbey left for an extended tour of Europe, Asia and Egypt. He clearly was trying to avoid being a witness in the upcoming patent battle between Libbey Glass and Owens Bottle. The same year he gave the museum $850,000 to enlarge the building. The museum was to be doubled in size to accommodate the thousands of articles in storage. In early January, while he was in Europe, the new South High School was renamed for Libbey. Libbey was honored and sent a check for $35,000 for a new athletic field. Libbey indicated he would like to go further in education by adding a technical high school and even possibly an endowed university in Toledo.

Libbey was spending most of his time in Ojai, redesigning the town as a resort and pouring in millions of dollars. He created local committees to address various parts of the plan, in the same way that he involved the Toledo community in building the museum. Libbey was president of the board of directors. The main streets and shops were converted to a mission-style arcade. Libbey purchased land for an inn, park, spa, and golf course and was particularly interested in the country club and golf course. He hired golf architect George C. Thomas, who also designed Rivera, Bel Air, and Los Angles Country clubs, to design a panoramic golf course. The course soon became popular with Hollywood movie stars. During World War II, the army stationed troops on the course, and Thomas's favorite holes — 2 and 5 — were obliterated. They had originally been designed resemble #3 at Pine Valley. Today they have been restored. Wallace Neff designed the clubhouse and won an award from the American Institute of Architects.

In his 1924 trip to Europe, Libbey was freer than ever before to focus on art collecting and study. He demonstrated the same intensity as in any endeavor. In his tours of Spanish galleries, Libbey was struck by the work of Francisco de Zurbaran (1598–1662). Zurbaran was a contemporary of Velázquez, whose art Libbey had collected. When he tried to find a Zurbaran, he found they were all owned by the state and the church. Libbey persisted and finally tracked down a Zurbaran painting that had been removed to England in 1839. He was able to purchase the *Flight of the Holy Family*.

December of 1923 would be one of the darkest months for Edward Libbey and signaled the end of Camelot for the glassmaker. In his mid–60s, Mike Owens was pushing on with projects such as automotive safety glass and new improvements to his bottle machine. Times were very good for Libbey-Owens and Owens Bottle. Owens had called a board meeting of Owens Bottle two days before Christmas, while Libbey was in town. Directors Libbey, Blair, Boshart, Biggers, and William Ford had gathered in room 1401 of the Nicholas Building to hear Owens lay out his capital needs for a bottle machine development project. After an hour and a half, the men took a break and Owens went to Biggers's office to make a phone call. Biggers would later note that there was a "fairly heated argument" before the recess. Owens and collapsed in the hall. He was carried to the leather couch in Biggers's office. He had been diagnosed in 1917 with heart problems. He asked for Monsignor Anthony Dean, pastor of Toledo's Rosary Cathedral. A secretary called neighbors and friends doctors Otto Sasse and Will Gardiner. Owens's family was also summoned, but arrived too late. The doctors could do nothing, and Father Dean gave Owen's the last rites. Libbey watched in tears as his old friend and corporate war horse died. He stayed until ambulance attendants took the body, struck with grief, realizing an era had ended. It was the end of a friendship and a partnership.

The city of Toledo mourned a great friend. Many residents came forward to describe of how Owens had given them money to help with problems. The funeral was one of the largest the city had known, with Bishop Samuel Stritch (a future cardinal) of Toledo and Bishop O'Connell of New York (an old friend) presiding. A

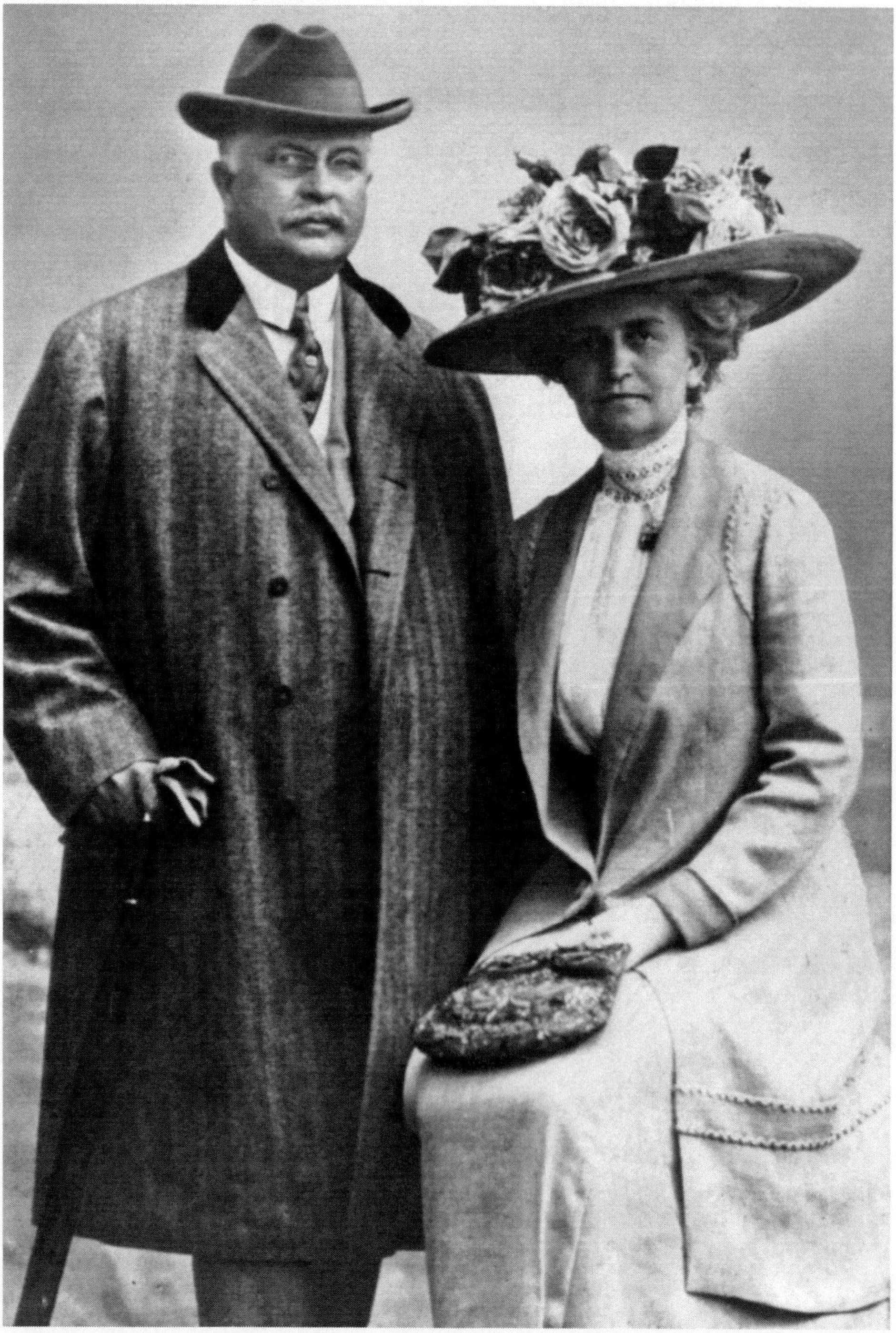

Edward and Florence Libbey, 1891 (courtesy of University of Toledo, Ward Canaday Center, Libbey Glass Records).

priest for whose education Owens had paid delivered the sermon. His pallbearers were union officers, Tom Rowe and William Clarke. Other pallbears were old friends. Libbey was named as an honorary pallbearer. He published a personal tribute noting:

> I believe that the name of Michael Owens will stand out as a pronounced example of what can be accomplished by vision, faith, persistence and confidence in one's creativity. The development in the life of my friend, Mr. Owens, has been an interesting experience. I am indeed proud to have a factor in his business career and to have been able to see the expansion of his ability, to have studied his thoughts, to have interpreted his emotions, to have had an insight into some of his secret thoughts and visions, all of which could not fail to impress me with his greatness of mind, ability, child-like simplicity, and devotion to his friends.

Libbey had captured the essence of Owen's personality, and it had been Libbey that drew the greatness from him. With Clarence Brown and Owens gone, Libbey had lost his two most valuable employees.

After spending a few weeks helping the businesses adjust to the loss of Owens, Libbey returned to his home in Ojai. He was now determined to take his leave from active management in his companies. He hoped to spend more time on golf and fishing in California. He was still deeply involved in the development of Ojai as a resort paradise. He was also planning a final grand tour of Europe and Asia and was spending time with the Toledo Museum. The death of Mike Owens made the city of Toledo appreciate both Owens and Libbey. In May 1924, Libbey returned to Toledo to receive the Rotarian Silver Cup Award for service at a lunch held on May 23. Speeches were given by John Willeys of Willeys-Overland; Blake-More Godwin, museum curator, Charles Feilbach of the Board of Education; John Cowell of the Chamber of Commerce; and Alfred Koch of LaSalle Department Store.

In the spring of 1925, Libbey remained president of his companies, but he was slowly backing out of the duties. In May of 1925, while staying in Paris, Libbey wrote the following letter to the annual meeting of Owens Bottle Company:

> You and I have reached the period of the end of the first stage of developments of the Owens invention, a most glorious period of pleasant and happy experiences, and one in which we may all feel a pride. Its conception, growth, and success cannot be surpassed, and I feel that in so far as I am concerned, I have left a legacy so rich in future prosperity to make the property of The Owens Bottle Company one of the greatest in industry. As previously stated at the last meeting of the Board of Directors at which we were all present, I feel that in duty to myself, I must decline to assume any official position in the company, and in this I am irrevocably determined. Therefore, I would ask you to convey my decision to the stockholders and my request not to elect me to any office.[1]

The board accepted the resignation but made him chairman of the board. By October, Libbey was reportedly longing to be back as president.

Libbey left Paris for his last collecting trip to Egypt in 1925. For years, he had been a supporter of the Egypt Exploration Society, and in 1924 had become the largest supporter of an excavation of Tell El-Amarna. This excavation was of the royal city of pharaohs Ikhnaton and Tutankhamen, who lived in the fourteenth century B.C. Libbey was invited to visit the excavation in 1925. He lived on site for several weeks at the staff house and worked in the field. This was one of the high points of his many trips overseas.

In late August of 1925, the Libbeys returned from a collecting vacation in Europe and Egypt. They spent the month of September at the Walbridge summer home in Pike, New Hampshire, moving on to Toledo to stay at their suite at the Secor Hotel. The house in Toledo was not being used and had no staff. Libbey often lived at the hotel while his wife was in California. He took a few short business trips in late October and early November to New York and Charleston. The Libbeys returned to Toledo and their hotel suite on Tuesday, November 3, 1925. They planned only a short stay before going on to their California home. On Friday and Saturday, November 6 and 7, Libbey worked at the museum with Stevens. He toured the new addition, two-and-half acres of floor space. He said he wasn't feeling well, believing he was catching a cold. The weather was wet and raw.

Many details of Libbey's final days are known through the memories of Secor Hotel room clerk Henry Sprang.[2] By Sunday morning, Libbey complained of not feeling well, and coming down with a cold. His friend and West End neighbor Dr. Otto Sasse was called to examine him. By Monday, another physician, L.C. Gosh, was brought in from Cleveland. Libbey's condition went back and forth through Thursday, when his fever returned. It was now clear he was suffering from pneumonia, and his bad heart made the situation bleak. Things got worse on Friday, November 13, as a new problem emerged. The hotel and city were under siege by Ohio State fans. Loud and drunken students were everywhere. The police were notified of the situation. The students could not know that a great industrialist lay dying. The OSU-Michigan game was a major event, and Ann Arbor had limited hotel space, requiring the OSU crowd to book rooms in Toledo. It wasn't a particularly good year for Ohio State, but they had just beaten Indiana the weekend before, and Michigan had lost its first game to Northwestern. The partying included. A concerned hotel manager called the Toledo police chief and the police blocked off the hotel from traffic and people. Only customers were allowed in the hotel, and police were placed on each floor. Nonetheless, Libbey died on Saturday morning, November 14, 1925, at 9:15 A.M. His death brought to an end an era in the glass industry. It was also the end of an era for the city of Toledo, which was known throughout the world as the Glass City.

Libbey's death brought peace to the city the following Friday. High school football games were cancelled, and flags were flown at half-staff. The city schools held their own memorial services. Libbey's body lay in state at the museum before a private service at his Scottwood home. Dr. Allen Stockdale of the First Congregational

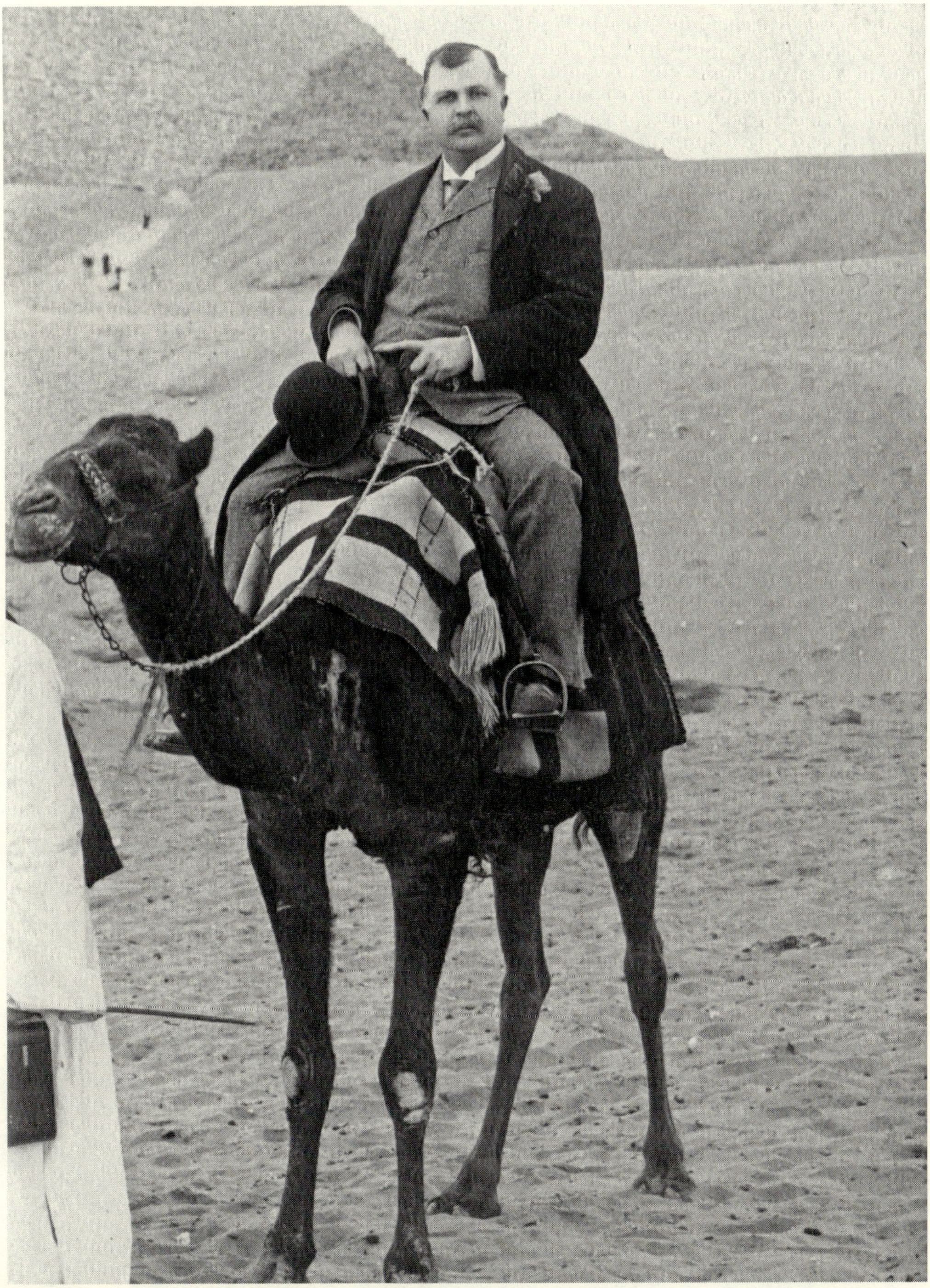

Libbey on an Egyptian art collecting trip (image courtesy of the Toledo Art Museum Archives).

Church conducted the service and Dr. Stevens of the museum gave the eulogy. Burial was at Woodlawn Cemetery.

Many sources have noted that the Libbeys both requested that all their private papers be destroyed upon their deaths. It appears that after Edward Libbey's death, his wife, lawyers, and friends did just that. They were clearly deeply private people. Family affairs, in particular, Libbey kept private. Some business correspondence survives in the archives of the museum and the University of Toledo, but only a few letters of a personal nature. One is to his and his wife's friend, Rosa Lang. This letter suggests that in writing, Libbey was much more open than his Victorian persona would suggest. He did enjoy beautiful women and liked to flirt, which can be seen in this letter to his wife's friend. The few personal letters of Florence. Libbey also suggest less formal relationships between her and the Stevenses. The museum has a single black box of personal items, which Libbey seemed to have wanted preserved. The contents include his batch book of glass formulas for New England and Libbey

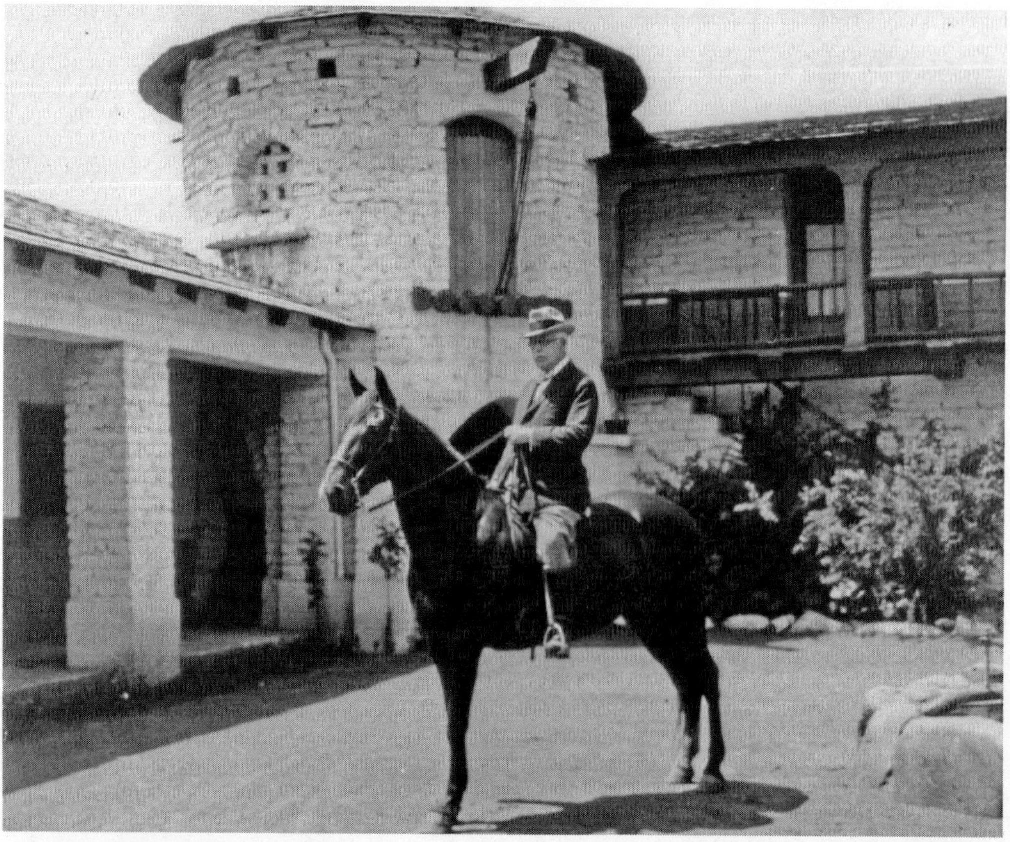

Libbey at his Ojai, California, ranch, 1890 (courtesy of University of Toledo, Ward Canaday Center, Libbey Glass Records).

Glass, as well as some business correspondence of his father. His preservation of his old batch book, long past its usefulness, is telling of how much he prized the chemistry of glassmaking and the craft itself. There is also a notebook with a few details of his trip to Europe in 1887. There are samples of the first product of the automatic bottle machine, which certainly represented a significant event in his career. There is also his last wallet and another personal notebook. This would suggest his wife added to or assembled the box after his death.

At his death, Libbey was worth $21,714,097. Most of the estate went to the museum, but Florence Libbey received $2 million directly and a trust fund of $1.5 million. The museum received his art collection and trust funds worth over $4 million and another $3 million for an art school wing and a music wing. After leaving major amounts to the museum and family, he left small amounts to friends as a sign of appreciation. His friend in Ojai, Harry T. Sinclair, received $25,000; and his friend and associate, George Stevens, received $100,000. Associates and friends received various amounts: Fredrick Geddes, $10,000; J.D. Robinson, $5,000; Solon O. Richardson, $5,000; William Donovan, $5,000; Charles Schemettau, $5,000; James Blair, $5,000. Many other friends received $5,000. These bequests emphasize Libbey's high regard of old friends and the importance to him of loyalty. Two employees, Robert Johns and Jack Flannigan, were singled out with $10,000 and $5,000 respectively. He left $100,000 to the Toledo Public Library and $300,000 to the Board of Education for teacher scholarships. In addition he left $15,000 to Libbey High School.

Chapter 19

A Model CEO

Libbey pioneered the role of company CEO, although that was not the term used. He focused on finance, investment, marketing, and community, leaving day-to-day operations to lower-level executives and managers. It was an approach at which many had failed. It is always difficult for the founder and owner to delegate. He must hire, develop, and trust his managers. And just as importantly, he must motivate them through ownership. Many of Edward Libbey's business practices can be traced back to those of Deming Jarves. His advertising strategies clearly go back to Deming Jarves and New England Glass. New England Glass was one of the earliest, if not the earliest, to go directly to consumers through newspaper advertising. The company was advertising in three newspapers in 1818 — the *Boston Commercial Gazette, Columbian Sentinel,* and the *New England Palladium.* In the 1820s, Boston & Sandwich Glass was also advertising in New England papers.

New England Glass had its own sales outlet in Boston requiring advertising early on. The sales outlet proved a better distributor than using many different shops throughout the city. Jarves set up factory outlets in Philadelphia and New York by the 1850s. Until the 1860s and the railroads, most glasshouses were regional in scope. New England Glass could readily move product down the Atlantic Coast to New York and Philadelphia. The mountains were a major roadblock to glass transportation since breakage in wagon trains was a big problem. There were only a few national producers such as Bakewell in Pittsburgh, which Deming Jarves studied extensively before opening New England Glass.

Libbey had many management models from which to draw. He and his father had worked with Deming Jarves, who was a pioneer in scientific management. Jarves brought creativity to industrialization. Intellectual property rights were key to this new management approach, and it was the job of the top operating officer to defend his intellectual property. Jarves's New England Glass was in operation prior to a strong enforcement of patents. The publishing of patents in many cases was too revealing. Formulas and pattern designs allowed others to make illegal copies. Patent law for pattern and designs was not fully established until 1861. Libbey went further than

Jarves in his legal defense of patents. Still, like Jarves, Libbey maintained many trade secret practices, as guilds had done for hundreds of years.

Just as important was the idea of corporate ownership of new processes and products. Jarves maintained the batch or formula book personally, a practice maintained by Edward Libbey. In the 1850s, many glasshouses were outsourcing their mold making, but Jarves held that the glass house should retain mold control. In the 1850s, most molds were made of cast iron. Independent mold makers, which were iron foundries, had access to the patterns and designs. Mold shops became sources of industrial espionage, but Edward Libbey maintained his own mold shop for Libbey Glass.

Libbey's defense of corporate intellectual rights was fundamental to the transformation of glassmaking from a craft to an industry. Libbey realized that quality and craftsmanship had to be maintained in the industrial model. Industrialization, more than anything, caused a change in the organizational structure from the factory floor to the owner's office. As the guild system broke down, a new management structure was needed. Libbey had to look through the period of struggle among craftsmen, foremen, owners, and unions, with no road map. He had to develop new management practices. He envisioned a type of industrial guild, but soon found that vision could not be realized. Furthermore, his multi-plant and product lines required a different type of top management. Most early owners in similar situations merely created individual companies that were independent, without the benefit of integrating shared functions. Jarves had done some integration, but Libbey's operations and products were vast in comparison. Only a handful of companies, such as Carnegie Steel and Standard Oil, operated at the same scale.

From the beginning, Libbey used the corporation as a legal and organizational entity. In the 1880s he had an advisory board of company experts such as Solon Richardson, Jefferson Robinson, and William Donovan. His second company, Toledo Glass, had Libbey and Owens as the experts in glass and employed a businessman, James Bell, and two lawyers — Clarence Brown and Fredrick Geddes. Their legal expertise would be the core of most of Libbey's company boards. They worked am intellectual property, leasing, patent defense, and anti-trust issues. Libby's board compositions were like those of many manufacturing boards today. He resisted adding bankers to any board, as he preferred to raise capital through stock instead of through bank loans. He focused on organization from top to bottom.

In addition to following the lead of Jarves, Libbey admired the organization skills of Andrew Carnegie.[1] Libbey had often read the hagiographic story of the Carnegie Steel Company — *The Inside Story of the Carnegie Steel Company.* He was even in the process of preparing a similar account of his organizations. Libbey and Carnegie both used stock to motivate their managers. Like Carnegie, Libbey built an organization to execute his policies. Like Carnegie, he maintained a partnership to hold various companies together. Unlike Carnegie, who had suffered a rebellion in the partnership, Libbey formed sub-partnerships so no individual or group of individuals could threaten the whole entity. Libbey Glass, Toledo Glass and Owens Bottle, for example, were separate organizations with few links other than Libbey himself. Legal battles between the companies demonstrated the autonomy they held. The loosely-bound Libbey empire was able to avoid anti-trust action by the government and still function as a monopoly, controlling prices. Libbey had integrated the ideas of Carnegie and Jarves into a very successful business strategy. The Libbey organization was as unique as those of either of the others. It was an unusual blend of autocracy and democracy. Libbey was able to maintain financial control but had to sacrifice some managerial control. However, this suited Libbey better, as he gave his managers far more freedom and autonomy than Carnegie.

Libbey, unlike Carnegie, used the corporation model instead of the partnership model. Still, he ran his empire like a partnership, serving as president of all of his companies. With the exception of Libbey Glass, a small group including Libbey, Clarence Brown, Fredrick Geddes, Mike Owens, and William S. Walbridge controlled the core of the companies. Later, John Biggers and William Boshart joined the group. This group behaved as a close-knit partnership, with Libbey in complete control. Such control allowed him personal freedom to pursue other interests and remain in strategic and even tactical control of operations. He could count on absolute loyalty. His older partners, of Solon O. Richardson, J.D. Robinson, and William Donavan, were kept from participating in the other companies because he didn't trust their loyalty. Richardson and the others had shown too much independence in some of the early decisions of Libbey Glass. Libbey had come to trust his lawyers, Brown and Geddes, and they would prove themselves throughout Libbey's career. The gang of four (Owens, Geddes, Brown, and Walbridge) would question, Libbey but never put up a direct challenge. Had Carnegie had such loyalty, the steel industry would be much different today.

Brown and Walbridge were Libbey's close friends as well. When he was in California or Europe, he could count on these two to follow orders. Brown had always been Libbey's sounding board on internal company problems. He was also a very skillful lawyer and executive. Brown was rewarded with stock and options over the years which would make him the second largest shareholder in Libbey's overall empire. Walbridge, of course, was family and was close to Owens and the pulse of the middle managers of the company. Walbridge also increased his stock holdings in the various companies as a reward for his loyalty. In general, it was a balanced group, two lawyers, two glass men, and a general businessman. Libbey understood the glass industry best, and Owens was the technical and operations man.

Libbey had many friends such as Stevens, Ashley, and Secor that shared his love of art; these he added to the museum board of trustees. Libbey was a true pioneer in applying the corporate model to a community museum. He welcomed other lawyers of the Brown firm, such as Geddes and Charles Schmettau, to the museum and corporate boards. After Brown's death, Libbey found a successor in James Blair, who was groomed as Brown's replacement by Brown and Libbey. Libbey was always charming and diplomatic in his personal dealings. He was slow to analyze a problem, but very methodical in his implementation.

Libbey was one of the first big business owners to fully understand the importance of his executives and managers in the overall performance of the company. He was unmatched in the recruitment and development of executive talent. He valued the creative mind and managerial skills. Libbey saw men like Owens, Richardson, Bock, and others as key parts of the organization. Over the years, Libbey had many "understudies" for Owens, but, they were difficult to maintain. Libbey understood that he needed backups for executives and key managers. Again, few owners at the time had such a comprehensive understanding of their organization. He groomed LaFrance for years as an understudy for Emil Bock. James Blair had been hired as Clarence Brown's understudy, which made for a smooth transition on the death of Brown. This use of understudies allowed the organization to absorb the loss of key individuals. Libbey understood the financial value of his key executives and was one of the first industrialists to take out insurance policies on them. This organizational reinforcement allowed Libbey to have more control and personal freedom.

Another part of Libbey's strategy was to avoid bank loans. In the earliest days he did receive help from local bankers such as Ketcham, but he always got a personal loan rather than a corporate loan. Libbey had experienced the problems of bankruptcies and bank foreclosures, and he was determined never to allow bankers into the corporate mix. He used family and personal loans often, but avoided formal corporate loans. He was also reluctant to pressure his friends to invest. Men like Brown came to his board as advisors, with little or no stock initially. Some historians blame his aversion to bankers on bankers' opinion of Libbey — that he depended on intuitive judgment versus detailed analysis.[2] This is unlikely; Libbey showed a passion for detailed analysis, and he had strong support in men like Brown. Others say banks saw Owens as being too risky, but again Libbey's companies after 1894 were solid and found it easy to raise capital in the equity markets. The fact is Libbey proved to be a brilliant financial manager, as did Brown. He probably drew criticism from banks because bankers were not asked to join the boards.

Like Carnegie, Libbey disliked conflict and dealt with it behind the scenes. Carnegie had a radical, maverick plant manager in Bill Jones, and Bill Jones made the early Carnegie Steel profitable. Libbey had Mike Owens to deal with. Owens was integral to Libbey's success, but his un-conventional approaches made him a difficult organizational fit. Bill Jones was the son Carnegie never had, and Mike Owens was the son Libbey never had. Both company presidents found it hard to refuse any request of their radical young managers and had to deal with the fallout. This weakness of Libbey would become his strength as a CEO. Libbey became a master of his organization, fitting people to his projects and operations. He allowed for internal competition, which at times was messy, but always profitable. Libbey became a model for CEOs. He remained strategic in his organizational design and built layers to insulate himself from tactical conflicts. In the end, Libbey would prove more skillful than his idol Carnegie in staying out of day-to-day operations. Libbey never fully controlled Mike Owens, but he learned to live with him.

Libbey's strength in incorporating Mike Owens in the organization bordered on a weakness at times. It would be easy to see Libbey as indecisive, and he could be overly thoughtful when faced with a problem. He struggled with major decisions, like the one to leave New England. He certainly would have been a poor field commander, but a CEO is not a field commander. Libbey continued to allow Owens to strain the organization to the breaking point. In fact, in 1920, two of the companies of which he was president were suing each other (Libbey Glass and Owens Bottle Company). Libbey's neutral approach cost him many friendships; in the end, Mike Owens felt isolated. Libbey acted to censure his executives only as a last resort. Libbey understood that creativity often comes with organizational strain. Still, despite the in-fighting, Libbey's companies continued to be very profitable. Today we might find a similar example in Apple Computer over the last few decades, where creativity and organizational stresses co-exist.

In the end, Libbey always seemed to make his decisions for the good of the company and his income, rather than for friendship. In many ways he is the boss we all want, allowing freedom and judging on results. Problems arose only when managers pressed for political advantage or a final blow against an opponent. In these final stands, Libbey might seem excessively cold, but that fails to account for the years of tolerance that often preceded these points of crisis. Some executives might have eliminated Mike Owens years earlier and had a more unified and peaceful organization, but they would have lost breakthrough innovations and millions of dollars in profits. One might look to Thomas Edison's inability to tolerate genius, and the eccentric Nicola Tesla at Edison Electric. Tesla went to Westinghouse Electric, which then surpassed Edison.

When we look today at Libbey's various interconnected companies, we see he ran on the cutting edge of organizational design. Today we call it entrepreneurial. He allowed creativity and drive to circumvent organizational structure. It was a dangerous approach, but it allowed individuals to rise to their highest level. It drove organizational-minded people like Solon Richardson crazy. But balance and long-term harmony depend on the person at the top. Libbey saw his role as one of leadership. He was to guide, not direct. Andrew Carnegie had used the technique to build a steel empire, but when he lost focus, the organization broke up. Such organizations require the personal leadership of an entrepreneur. The creativity of such organizations can be powerful. Libbey made breakthroughs in flat glass and bottle making and in many revolutionary processes. In the community, he built a new model to organize and finance museums. More recently, we have seen the same success by Steve Jobs with Apple Computer.

One area in which Libbey surpassed Carnegie in was the use of cost accounting for making machine and process decisions. Cost accounting was key in Libbey's research and development efforts. Cost analysis led him to license some machines instead of selling them. More importantly, Libbey made cost accounting part of his sales campaign. In 1905, Owens Bottle issued a brochure, "Comparative Estimate of Production Costs." It compared hand blown bottle costs to automated bottle production to help sell the automated bottle machine. It was this cost accounting approach that helped focus the research effort of Toledo Glass, and helped him revolutionize the machine tool industry. He would lease machines and base his royalties on the savings to the customer. This allowed the customer to get the machine without high initial capital outlays. Most machine tool companies wanted to sell machines. Libbey was giving companies a virtual industrial credit card, which paid very high interest rates. He allowed payments to be made in machine-generated savings. Cash-short companies were allowed to pay in stock. Interestingly, years later when the first industrial robot manufacturers had problems selling their new machines, they used the Libbey strategy to lease, based on labor savings.

Another overlooked part of Libbey's executive style was his use of the committee system. He had initiated a board of directors at New England Glass and Libbey Glass. The board tackled tough decisions but was advisory in nature, as members did not own stock. The committee system was the basis of leadership in the glass unions at the time, and Libbey emulated it and integrated it into the corporate structure. Libbey's boards played the role of the President's cabinet. In many ways, his approach was like the Carnegie partnerships. Libbey retained control over his individual companies, allowing argument to a point. A tight network of friends and associates controlled stock allowing Libbey unquestioned control. However, he didn't rule as an autocrat but listened and processed information. He could be slow and deliberate. Libbey's powerful use of the committee system allowed him to focus on the key points of the business, while giving him much personal freedom.

Members of Libbey's boards were advisory in nature, but also working corporate members. Unlike corporate boards today, true management belonged to Libbey. The boards were like a general's council of war. These working committees were filled with working managers or major stockholders of the company. Some, like Clarence Brown and Fredrick Geddes, were from outside the company and merely had business ties, but Libbey quickly involved them and they became major stockholders. Libbey had used subcommittees often, to study particular issues. The museum effort was fully managed by committees. The system allowed Libbey to be involved with a seemingly endless number of projects, while wintering in California and summering in Europe.

The system depended on the careful selection of directors. Libbey maintained stock control and, in effect, was in charge of the selection of directors. His directors were highly motivated and rewarded. A director had access to privileged inside information on investments. Men like Clarence Brown, Fredrick Geddes, James Blair, and Mike Owens became very wealthy through the investment opportunities offered. The boards had shared directors, but the only person who could step into Libbey's role was Clarence Brown, and he did not even serve on the Libbey Glass board. Libbey trusted family members like W.S. Walbridge, who together with Brown followed orders and gave honest feedback. He could also count on the loyalty of Mike Owens and George Stevens at the museum. Clearly loyalty was an attribute that Libbey searched out.

History understated Libbey's role in the success and growth of his empire. Many erroneously point to the accomplishments of Michael Owens as being more important. Owens had a set of skills that were critical to the development of the bottle machine and flat glass machine, but Owens could not have functioned inside a corporation without Libbey. A reporter in 1919 looked at this unique relationship:

> Together they attained success because Owens possessed, in addition to his mechanical and glass working genius, practical ability for factory management, and good business sense; and because Libbey possessed, besides capital, a rare financial and administrative capacity, unfailing judgment, and a vision of the results which would follow the proper development of an idea. Moreover — and this is quite as significant — both recognized their own limitations, had full confidence in each other and so could coordinate their efforts. Owens will tell you that the company's great success is due to Libbey. Owens will tell you that the company's great success is due to Libbey.[3]

Clearly it was Libbey, not Owens, who was behind the corporate success. Owens lacked even the most basic administrative skills.

The key to the pair's success was Libbey's personal financing of Owen's risky projects, especially with the bottle machine. He was even hesitant to involve his friends in taking a risk. Libbey had a great faith in Owens's ability to deliver results and could goad him to achieve more. Always wanting to please Libbey, Owens would rise to the occasion. A few years before his death, Libbey noted: "Mike is a remarkable person. He could make a success of anything. If he'd taken up the law he would have been a successful lawyer just as he has become a successful inventor and business manager."[4] Libbey was the only person at times that believed in Owens. He saw Owens less as an inventor than as a project manager. Owens could push a project to a finish. It was hard to categorize Owens, but he delivered on goals. Libbey understood this.

Libbey created a system of checks and balances in his organization. It was a personal network to be sure, but it was successful. Libbey, for example, brought his brother in-law, W.S. Walbridge, into Toledo Glass management to check Owens's propensity to spend money. Walbridge tightened up the purse strings, reporting to Libbey formally and informally. Libbey often used Clarence Brown as another curb in many of Owens's far-reaching projects. This was the same system that Andrew Carnegie had used among his partners, more of a political methodology than a management strategy. Libbey's method would probably not suit most business managers, but for Libbey it worked. It gave him absolute control of the organization, but it required loyal lieutenants. A network approach instead of a linear management hierarchy. The network structure is hard to hold together, but can lead to creativity. Solon O. Richardson, who firmly believed in a formal staff and line organization, had trouble with it. Libbey could be tough when loyalty was breached, as when Richardson, Robinson, and Donovan argued strongly against involvement in the 1893 World Fair. They were frozen out of many future investments.

Libbey's administrative and executive skills were probably best exemplified in the successful Toledo Museum, a community museum, a financially independent, functioning unlike any other. Libbey contributed the most to the building fund, but by the second decade of the century, the museum was fully supported by its members with no endowment or municipal aid. The museum initially lacked significant funds to purchase art, but still was an amazing accomplishment built through committees and community networking. Libbey broke new ground with his unique fundraising. He used advertising and networked with community groups beyond those interested in pure art. Libbey went beyond other philanthropists like Andrew Carnegie, built over 500 libraries, requiring the community to supply the books. Often the libraries failed decades later. Libbey created a living and functioning organizational legacy in the Toledo Museum. The museum remains, while many of Carnegie's libraries are gone or in a state of disrepair or have been taken over by government. It was a new and very different type of philanthropy that some call "constructive" philanthropy. The Libbey sense of community can still be seen in the museum today, especially in its mission of education in its programs and offerings.

Libbey's approach to developing a community museum gained as much international and national recognition as his business triumphs, if not more. The world and nation came to Toledo to see the museum's operating success. Steel man Joseph Butler of Youngstown modeled the Youngstown Museum of Art after that of Toledo. Butler would spend weeks with Libbey and Stevens discussing organizational structure and community networking. Paul Rea, secretary of the American Association of Museums, noted: "The history of the Toledo Museum is a most conspicuous illustration of the response which the people will make to the advances of museums. In a few instances have these advances been made so earnestly but where they have the response has been as ready in proportion to the resources of the community." The difference, of course, was the community organization that Libbey had created. In this respect, Libbey's committee system rivaled that of Rockefeller's Standard Oil committee management system. Libbey's museum committee members included industrialists, artists, politicians, educators, and residents.

Chapter 20

An Industrial and Labor Republic

Edward Libbey lived through great changes and turmoil in the glass industry, which was finding its way from a handcraft to the assembly line. Glass products fell somewhere between art and consumer goods. Glass workers fell somewhere between craftsmen and factory drones. Many times these craftsmen of old opposed both industrialization and unionization. Unionization was a setup down in status and freedom, yet there seemed to be no choice but to unionize. Libbey loved craftsmanship but had to industrialize to survive. Communities which had been part of a labor republic in the crafts system were now cities of industrial workers. Libbey was torn between manufacturing art and manufacturing consumer products, but like his workers, was committed to making the conversion to industrial manufacture with dignity. They all wanted the status due craftsmen, but also a piece of Middle America. Libbey never viewed himself as an industrialist or even a capitalist but as a manufacturer of art and an employer of craftsmen. He wanted the simplest of his workers to understand and love art. Even as he automated processes, Libbey wanted to maintain the craftsmanship of the product. He believed that a worker's republic was possible and resisted the new industrial order He was clearly a bit of a romantic and an idealist in a manufacturing world that ignored the value of craftsmanship. While industrialization won, Libbey was able to ease its pain on the community and give hope for a new vision. The Toledo Museum stands as a history of craftsmanship and its relevance in a manufacturing world.

Libbey's career highlights the successful transition from crafts production to industrialization. His workers were well paid and maintained a dignity in the community, characteristics lacking in other industries of the period. He avoided the labor problems of the steel and other manufacturing industries and incorporated much of the crafts and guild systems into the new workplace of industrial production. Libbey even created new positions, such as designer, in the industrial system. He allowed for upward mobility in the workforce, a holdover from in the crafts system. Libbey's monopolistic practices of price control and limited production favored an easy transition of the craftsman to industrial worker. He offered craftsmen the new, skilled industrial jobs at good wages, unlike the steel industry where the transition from craftsmen to industrial worker was a humiliating experience of job loss to low-wage immigrants. Libbey saw the glass worker as a skilled worker.

To understand this change in the glass industry and its communities in the last half of the 1800s, one needs to look at the crafts system and its relationship to owners and the community. In the crafts system, owners and craftsmen shared management decision-making. The relationship between management and skilled workers was one of cooperation, not coercion. It wasn't perfect, but there was respect on both sides. Libbey, in particular, was in awe of the craftsman's ability to create beauty. Likewise, the community and factory saw each other as interdependent. This community relationship had evolved in America's first factory town, Lowell, Massachusetts, in the 1820s. It carried over to cities like Cambridge, that adjusted school schedules to help the factory and allow boys to gain crafts experience. Craftsmen held status in the community. This stood in direct contrast to Carnegie's Pittsburgh, where immigrants lived in great slums. The craft relationships were not lost in the new industrial order of Libbey Glass. The Toledo Business Men's Association understood the relationship of local manufacturing and the community. Libbey understood it in his desire to have good housing, schooling, and recreation available to his workers in the move to Toledo. Libbey and the Toledo businessmen realized the Industrial Revolution was changing things, but the end result did not have to be the industrial slums of Dickensian's England and Carnegie's America.

The New England glass towns were craft communities. The companies and towns worked in a paternalistic partnership. It was not, as many suggest, a type of communism. The company built the workers' houses, but workers paid the company back on an installment plan. In many cases such as at Sandwich, the company had a store for food and supplies which was often very profitable. Still in times of unemployment, food might be given freely to families, credit extended to workers in hard times. The town and company were truly a family. Injured workers were kept on the payroll doing clerical jobs. Widows were often given coal to heat their homes. Men who started as boys worked into old age. No man was ever discharged for age or disability. The community adjusted school schedules to allow boys to work. The company helped the community by supplying a fire department. Boys would visit houses, notifying the craftsmen of the status of the glass pots.

Boys would receive free education at company-sponsored schools and crafts programs. Libbey's approach to child labor was much different than those in the pressed glass industry. Boys in these production-oriented plants were becoming a pair of hands in the 1880s. The jobs were no longer the first step in learning a trade. Libbey avoided the orphan trains of the western glass industry. While he hired boys, they were often able to progress in one of the crafts such as gathering, blowing, or cutting. Libbey promoted and supported manual training schools in the community as well.

He never saw boys as cheap labor; neither did his managers such as Mike Owens. Libbey saw those youngsters as future craftsmen and supported their education, including in industrial arts. Libbey formed industrial guilds based on art. These guilds were industry-related and would meet weekly at the museum. He foresaw the laborers and workers playing an interactive role with the museum, in his vision of the industrial guild and labor republic.

One of the overlooked factors in Libbey's decision to move to Toledo was this vision of an industrial republic. Toledo of 1888 was still small. Libbey envisioned an urban crafts community in Toledo, where could have his own suburb yet be part of a larger industrial city, as East Cambridge coexisted in its big neighbor, Boston. Libbey Glass could be the core of this growing city. The famous welcome and parade was community reaction that Libbey wanted but would never have gotten in a city like Pittsburgh. As he had envisioned, Libbey became the industrial heart of Toledo. Certainly, the work of his in-laws in the development of the manual trades and of skilled craftsmen and charting their role in society impressed him. Toledo had a manual training high school located on Adams Street, which could be a source of younger workers and boy workers. Toledo's Tri-State Business College was a well-known school for accountants and business training.

Toledo did grow in many ways in accord with Libbey's vision. The Toledo Museum of Art was more than a gift to the city; it was a symbol of Libbey's industrial republic. The museum represented an educational network to teach children the arts, with close ties to the public school system to achieve this mission. Libbey had given freely to the school system as well. He hoped to bring the arts not only to the rich, but to the sons and daughters of the workers. Libbey considered skilled labor on a par with professionals work.

Libbey was a Republican and a protectionist. These were considered fundamental to the vision of an industrial republic. Like President William McKinley, Libbey believed that the industrial republic was to be an alliance of management and labor. He was no robber baron; he represented the ideal of the Republican protectionists. The Democratic opposition had always feared that tariffs would only line the pockets of the owners. The industrial republic needed the profits of tariff protection to be invested in new factories and product development. Libbey, more than any other American capitalist, represented the ideal. Motivated by stable markets and increased profits, Libbey invested in the future of the industry. The development of the automatic bottle making machine and the flat glass machine were the result of Libbey pouring profits into development. Toledo Glass Company was an investment in research and development that revolutionized the glass industry. Glass tariffs created jobs, lowered prices, and benefited the community. Libbey's view of an industrial republic was a partnership among the government, unions, the community, and companies.

Libbey believed in American exceptionalism, as did McKinley, and his protectionist views evolved from this belief. Tariffs were a key part of his success. He was competing in a global market and needed tariffs to be competitive. European wages for skilled craftsmen were 60 percent to 70 percent less than those in America. An unskilled day laborer from 1880 to 1910 in the United States averaged $1.40 a day, a boy averaged 50 cents a day. But a European laborer made 15 to 30 cents an hour. The official estimate was that wages were 50 percent to 25 percent less for European workers in 1882.[1] Since labor at the time was the major cost in the glassmaking process, these differences were very significant. Ohio Republicans such as James Garfield and William McKinley had served as chairmen of the Ways and Means Committee of Congress. Glass tariffs varied by product, but generally were 40 percent to 60 percent on the value shipped during the Libbey period, which supplied fair protection to the American worker and owner. Libbey, of course, was a contributor to and a supporter of these Republican politicians. It was a partnership. Libbey was McKinley's ideal—an industrialist who poured tariff-driven profits back into the business and jobs. Much of McKinley's opposition arose from the fear that tariff profits would line owners' pockets.

Democrats feared high prices and owners not investing in plant expansion. Libbey proved them wrong on both counts. First, Libbey invested aggressively in innovation, plant building, and equipment. Between 1890 and 1905 he achieved a labor republic. Wages increased absolutely and when measured against the cost of living.[2] Employment in glass and other major industries increased dramatically, creating a labor shortage that spurred immigration. The stability of the glass industry allowed long-term research and development, which paid off. The great technological advances of Toledo Glass pushed the price of glass products down dramatically. The price decreases created vast new markets for glass, and the industry and employment grew. More than in any other industry, Libbey's glass companies demonstrated the success of the labor republic. Libbey supplied the responsibility, patriotism, and far-sighted vision to make it work. However, McKinley's Tariff Commission constantly reviewed tariffs, prices, wages, and employment. Where many capitalists did pocket the profits from tariffs, Libbey invested in research and development, plant expansion, and capital investment.

Libbey showed that by stabilizing the markets, owners could invest in research and development. The biggest restraint on capital investment in industry is fear of declining prices or profits. Domestic production accounted for nearly 90 percent of the U.S. glass market from 1880 to 1900.[3] Today we see falling prices and lost market share due to globalization, and this makes domestic capitalists hesitant to make major investments in research and development. In Libbey's industrial republic, automation resulted in more jobs, not fewer. The bottle machine, for example, reduced the price of bottles to the point that large new markets such as beer bottles, milk bottles, and drug bottles opened up. While initially some blowers lost their jobs, in three years the overall employment of the glass bottle industry had tripled. Tariffs had done amazing things for the glass industry using Libbey's approach.

Libbey agreed with J.P. Morgan's belief in a peaceful and cooperative marketplace. Prior to the full implementation of the Sherman Anti-Trust Act, Libbey belonged to a number of price- and market-controlling pools. He tried to control the impact of his automation technology on the market price of bottles, creating market monopolies through his leasing arrangements. He helped transition skilled laborers to highly paid machine operators. Libbey, however, approached capitalism not as a businessman but as an economic philosopher. Like all capitalists, most of his business decisions were self-serving. He was shrewd but never ruthless. But he always stayed within the law, as did Libbey's friends such as Clarence Brown, Carl Spitzer, Fredrick Geddes, Aaron Chesbrough, Charles Ashley, and William Hardee.

The social structure of the industrial republic was also unique. It was a vision of skilled craftsmen and workers with time to purse the arts and education. Libbey imagined a worker villa like those he had seen in Europe. Unfortunately, to some degree, it was an idealization of the Industrial Revolution. Carnegie gave communities museums and libraries, but Carnegie's workers had neither the time nor money to fully appreciate the arts. Libbey envisioned the glassworker as the aristocrat of labor. This did not come to be, but Libbey's workers did have a better life than Carnegie's steelworkers. In the end, the Industrial Revolution slowly eroded the status of the glass craftsman until he was like another factory worker, but Libbey tried always to maintain his employees' dignity. Libbey understood that unions wanted to slow the process of industrialization. Libbey expected art glass and craftsmen of Libbey Glass. But his companies' automation and inventions eclipsed that vision. While art was being eliminated in the glass houses, Libbey tried to maintain its role in the community.

An art community made perfect sense in Libbey's view of a labor republic. An art museum would offer inspiration for the working class. It would improve their lives and worldview. The Rev. George Wallace of the First Congregational Church titled his museum inauguration address as "The Message of Art and Its Relation to Religion." Wallace noted: "The true mission of art is to increase the interest in the sublime and beautiful and that the great masterpieces are but the expression of a human life and that as a nation increases in civilization there is a corresponding increase in fine arts."[4] The *Toledo Times* further editorialized: "Toledo little realizes the magnificent work that is being done here by a little but ever growing circle of public spirited men and women in the direction of giving larger content to the spiritual life through the work of our art museum."[5]

Libbey believed that art appreciation would inspire both industrial arts and architecture. The *Toledo Times* supported that view in 1909: "People are beginning to look forward to a city beautiful, as well as a city commercial. We are learning to say of a building site, not that it is the most central, or that the most ground can be had for less money, but that a building or a group of buildings would be attractive in that place."[6] The museum would inspire architectural beauty in the city over the decades.[7]

Libbey had always envisioned the art museum as the educational core of the industrial community. As he said in Paris: "The history of art is the history of the human soul, in the development of its primary impulses.... This inborn yearning found in all people, needs the guiding hand of intelligence and opportunity."[8] Libbey was very different from most philanthropists in this respect. He didn't see upward mobility as Darwinian, as did Carnegie, or as Calvinist, as did many Eastern philanthropists. He saw it as environmental. Thus, given opportunities and education, members of the working class could rise to their fullest potential. It was a view consistent with the American concept of free education for all. With education, a community might expect to rise above social evils. While at times his lifestyle appeared elitist, Libbey was no elitist. To a certain degree, Libbey's love of art did separate him from many, although not from children and close friends. He regarded craftsmen as artists.

Libbey believed that art could upgrade society, as Carnegie believed that books could. It was not the paternal capitalism practiced by men like George Pullman, H.J. Heinz, and George Westinghouse. Libbey's paternalism was focused on the larger community, not just the workforce. He had lofty goals, but art would always be a hard sell in industrial towns. His workers were well paid and had good benefits. His managers always shared in the success of the company. Toledo did not have the industrial ghettos of Carnegie's Homestead and Braddock. Libbey might argue, as did George Westinghouse, that the best philanthropy was a high-paying job. Libbey's intense focus on his passion for art can be seen as self-serving. His other community philanthropy was reserved. Unlike Carnegie, Libbey never saw himself as destined by providence to change society. His mission of art education was an earthly goal. While he was a rich industrialist, he never wasted money on lavish parties like those of Fifth Avenue's millionaires of the period. Libbey lived extremely well, but he earned his money. In a very limited view, he probably could have given more back to the worker. Libbey was every bit the capitalist, and he gave back in his own way. He did so not out of guilt, but in a sense of responsibility to the community. Society needs its Carnegies and Libbeys to advance the arts and beauty. The Toledo Museum of Art continues to invest in the future of the community and continues to pay cultural dividends.

Libbey saw the link between culture and industry as cooperative and symbiotic. The Rev. Gustavus Bierdermann of Toledo, "conjectured that a municipality which invested money into the artistic and cultural life of the city was bound to attract a large population with many industries."[9] Such a relationship was commonly seen in Europe. Libbey had considered cultural factors in moving Libbey Glass in 1888. The city also realized that tourists would bring money into the community. Libbey had a deep interest in community development, which he also showed in his work with Ojai.

Chapter 21

Industrial Designer and Marketing Genius

Edward D. Libbey represents the nineteenth century confluence of the craftsman and industrialist. For him, that confluence was exemplified by the innovation in the glass industry of industrial designers, a job he pioneered through his business and the art museum. In many ways, Libbey had one foot in the eighteenth century and one in the nineteenth century. As a youth, he had admired the art of glassmaking and the glass craftsman. At the same time, he helped pioneer the industrial concepts of automation and division of labor. The demand for cut glass required more and more automation to stay competitive. Libbey's factory layout reflected a new approach to the old arts of cutting and engraving. For centuries, a piece of cut glass was the work of one major craftsman. A single piece could take days to complete. With cheaper glass and processes such as pressed glass came major cost pressures on the cut-glass industry. In the early 1880s, Libbey set up a kind of crafts assembly line, with stations for roughing, finishing, and polishing. The Libbey Glass cutting department of the 1890s resembled an assembly line of artists. It was a perfect application of Adam Smith's division of labor to art, with the industrial designer overseeing the work. Libbey maintained the artistry in the manufacturing process. He even developed the concept of cellular manufacture to produce individual art pieces. His famous punch bowl sets were primarily the work of one craftsman over a period of weeks. His factory had the same layout as Harley-Davidson today, which builds custom motorcycles in cells off its assembly line. Libbey's cellular approach allowed him to customize product when needed, maintaining his world quality-reputation. Customization and quality allowed him to compete with the producers of lower-cost pressed glass, who often had to sacrifice craftsmanship for production.

Libbey had a tremendous respect for the creative ability of the glassworker. Early in his career, like Jarves, he imported artists-craftsmen from Europe. His first trip to Europe to study needle etching had given him a real insight into the skill of these craftsmen. Libbey's best hire after Mike Owens was Joseph Locke, who built the company's early reputation in cut glass. Libbey hoped Locke as craftsman manager would be a management model for the industry. Libbey loved the art of glassmaking, and he never gave up on its incorporation into manufacturing. In many ways, his vision would evolve years later in the industrial design and art glass deco movements.

The start of Libbey's success came with Locke in 1882. Locke was an English glassmaker who had won gold medals throughout Europe, a prolific craftsman who wanted to further increase his output. Like Chihuly today, he used a team of craftsmen. Chihuly functions as the industrial designer." An engraved or cut piece under Libbey and Locke became a cooperative, creative effort among the various cutters and engravers. This freed Locke to create more by freeing his hands, but maintained quality. Libbey would expand this approach over the years with more division of labor. He encouraged craftsmen like Locke to be creative and innovative. Men like Locke flourished under Libbey's style of management.

Libbey's first success in the automation of art was in the application of needle etching to lamp globes and shades. That had been his first real management assignment in 1874. He was sent to England to look at newly developed lathes that could replace the tedious hand-etching process without reducing craftsmanship. Often one artist required three weeks to complete etching of a lamp globe. Using a lathe for needle etching reduced that to two days. The etched lamps were extremely popular, and Libbey was able to supply this huge market. It was a revelation to a young Libbey that the creativity of the artist could be manufactured. The design itself was the art, not the methodology of reproducing the design. He had demonstrated that art could be transferred by machines. More importantly, Libbey discovered that men and machines could cooperate to create art.

Libbey and Locke then applied this assembly line approach to making English cameo glass. Cameo glass was extremely popular in Europe, but its high price limited it to wealthy buyers. Cameo glass was art, with beautiful busts and scenes. Done by a single artist, it required cutting, etching, engraving and enameling. Locke was able to control the design, yet divide the crafting among several artists. This allowed him to create multiple copies of the same designs. The time for even a complex cameo could be cut by more than 50 percent. Libbey soon promoted Locke to superintendent, and he coordinated glass design throughout the whole process, from melting to finishing. This reduced costs while maintaining craftsmanship. Libbey applied the principles to his cut glass art pieces. By 1910, Libbey had an assembly line of 200 cutters. He distributed the work and divided the labor. Cutters went from rough to fine cutting on one piece, which moved down the line. The artist often did the finishing. Libbey's organization never sacrificed quality for quantity. He maintained a guild-type pay system, which reinforced the idea of craftsmanship. On certain orders he might assign a dedicated craftsman, as with many of his punch bowl sets.

Libbey always maintained that Libbey Glass was a company of craftsmen. While his bottle and flat glass operations adopted technology and assembly line production, Libbey Glass often preferred handwork for its cut glass pieces. The company had many artists in its employ. Libbey also collaborated with his friend, customer, and competitor, Louis Tiffany, and he never lost his love for the art of glass. He demonstrated this love in his glass collection, which traced the history of glass. Like Henry Ford, who always had one foot in the factory and the other on the farm, Libbey was pulled between manufacturing and art. He tried his best to align the two in the face of the Industrial Revolution. Like Ford, he was very much responsible for the advance of that revolution, yet he mourned the death of the crafts system. Libbey Glass would always stand for craftsmanship under Libbey's leadership. But even in his automated processes, Libbey demanded craftsmanlike-quality. Much time was spent on the Owens bottle machine in ensuing that its product rivaled the handmade bottle.

Libbey created beautiful art glass, but his success came in marketing it to consumers. While he advanced automation and production, he never lost his love of cut crystal. Libbey branded his products, making Libbey Glass synonymous with cut crystal, bearing beautiful cut trademarks. His use of product branding was unrivaled in the 1800s with the exception of fellow fair exhibitor, H.J. Heinz. A recent study of the history of marketing concluded that "few men better understood how perception of rank, notions of propriety, and conceptions of integrity affected consumers' choices than one of the principal architects of the cut glass boom and creator of a branded crystal that bore his name."[1] Libbey's strategy combined the desire for status, product quality, and reasonable pricing to build a vast middle class market for cut glass. His products succeed against the cheaper pressed glass tableware when other cut-glass companies

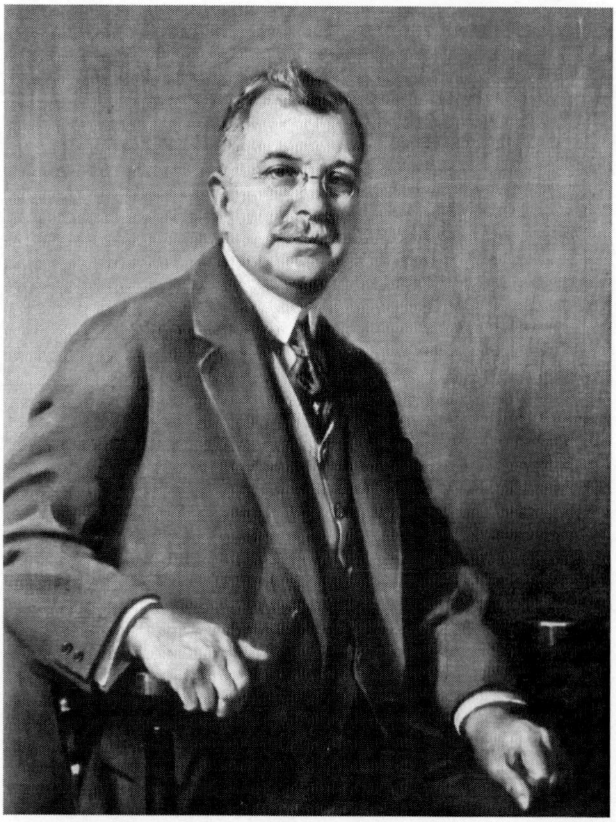

Edward Drummond Libbey, 1920 (courtesy of University of Toledo, Ward Canaday Center, Libbey Glass Records).

failed. Libbey looked at Tiffany's success in branding as the ideal. He never sacrificed quality to cut cost. He maintained the highest lead content of any manufacturer to assure the feel of quality.

Libbey's branding strategy evolved from many experiences, including his visit with his father to the Philadelphia Centennial Exposition of 1876. The importance of that visit can be seen in the fact that a copy of the Centennial program was preserved among his few private papers on his death. The launch of his strategy was his pavilion at the 1893 Chicago World Fair. Bankers complained that he overspent and over-advertised, but Libbey believed it was necessary to establish him as the world's greatest cut glass manufacturer. He paid a premium for a central location on the midway and for the exclusive sales rights for glass products. Clearly, Libbey intended to dominate the fair. He sent over half his workforce from Toledo to support the effort. He used the new electric lighting creatively to highlight his cut glass. A reporter at the time called it a "room lined with diamonds," and a recent historian said it "exuded a phantasmagoric glow, a spellbinding magic."[2] Another tactic that Libbey used at the fair, creating a sales boom, was personalization. Libbey had his artists acid etch a name on any glass piece purchased, and the public loved it.

He took the gold medal as planned, but more importantly, he followed up on the plan to make Libbey glass mean cut glass in the consumer's mind. Just as importantly, he wanted to redefine how the consumer looked at cut glass. It was no longer restricted to the upper class but could be readily purchased by the middle class. Through assembly-line division of labor, Libbey had brought the price into the range of the middle class, but now he needed to win over the middle class consumer. Libbey followed his fair success with a national advertising campaign (a first for glass pieces) using magazines such as *Women's Home Journal, Harpers Bazaar, Saturday Evening Post,* and *Cosmopolitan.* The advertisements highlighted beautiful table settings using cut glass that were within the budget of the average American. Libbey emphasized cut glass as a great wedding gift, an idea that seemed to resonate with the public. He pushed the concept of buying a single pattern for a whole household of glass pieces. Many of his advertisements focused on the art of making glass and its history. Medals won were often highlighted. The sales Libbey created bucked the economic downturn of the mid 1890s. He successfully teamed up with Louis Tiffany to produce silver and cut glass combinations.

Libbey also used the fair to launch and expand his sales network to reach the middle class. In particular, he expanded into jewelry stores and dry goods stores, along with department stores. He expanded his lines in department stores such as B. Altman, Siegel-Cooper, and Stern Brothers in New York; Marshall Field in Chicago; Boggs and Buhl in Pittsburgh; and Wanamaker in Philadelphia. Focusing on weddings, Libbey published booklets for retailers such as *American Beauty, Things Beautiful,* and *The Gentle Art of Giving.* Libbey found that jewelry store windows were an excellent way of advertising and projecting an image of quality. Libbey tried to lead or create a trend rather than waiting for the evolution of the American middle class. The notion of selling patterns was highly successful for five decades. He fostered contests among his employees to design new patterns it was common for some of his managers to have patterns patented under their names. Libbey's brand was as highly valued as Tiffany in many lines, but he also had captured the middle class. His branding was as successful as that of H.J. Heinz, who used a similar marketing approach.

Libbey tried to bring men into the cut glass market by promoting cut glass punch bowl sets as recognition tokens for a achievement. His cut glass punch bowl took the gold medal at the 1893 World Fair, and that was the cornerstone of his strategy. He widely publicized his punch bowl gift to President McKinley, and offered cut punch bowls as prizes at big events. They were often displayed in department and jewelry store windows in major cities. One set was given as a prize in the great race between an unknown Henry Ford and the world famous auto racer, Alexander Winton. Winton, a lover of Libbey cut glass, ordered the trophy himself, assuming he would win the race. Ford won the race and launched Ford Motor Company, but years after Ford's death, the punch bowl was sold at auction and it disappeared from view. It remains a mystical quest for many collectors to find it. Many believe the famous punch bowl sits in someone's house, full of fruit, or possibly at some little-known antique shop. Another famous punch bowl was made in 1902 for Toledo magnate George H. Ketcham. It celebrated Ketcham's world record-holding trotting horse, Cresceus.

Libbey went further in his concept of industrial design and the synergy between industry and art. He pioneered this concept through the Toledo Museum and later in its School of Design. Art, and in particular some hand arts, were the core of an industrial base, he believed. In the 1930s, these beliefs would lead to an industrial design movement in this country, but in the early 1900s, they were not fully understood. In 1907, Libbey supported the American Free Art League, whose main function had been to eliminate duties and tariffs on art coming into the U.S. The organization, however, went on to support art to help build industry. According to the American Free Art League:

> Germany has lost no time in founding technical schools.... They have revolutionized the industries of Germany. The second stage of the German government's task in building up German industries has been the erection of art museums where her designers might acquire skill and good taste from the study of the best examples of art. The government has spent millions of marks in building the great new museums of Berlin alone. It will not do for our government to adopt a less progressive policy if we are to compete in the world's markets.[3]

The theme of an industrial designer replacing the master craftsman was infused in his vision of the museum.

Libbey's vision is still celebrated, as noted in the museum's 2002 exhibition, "The Alliance of Art and Industry." The director of that exhibition noted:

> This exhibition demonstrates unequivocally that the successful collaboration between the museum and many of Toledo's industries during the first half of the twentieth century not only improved the lives of the city's inhabitants, but also significantly contributed to shaping the identity and lifestyle of every person's everyday life in modern America ... [and gives] important insights into Toledo's pioneering role and engagements of the professional designer in shaping consumer products and mass taste in this country.[4]

More than any earlier industrialist, Libbey stressed the need for industrial designers in post–Industrial Revolution organizations. Libbey had seen it in Libbey Glass, as he automated the process, eliminating the master craftsman. He needed some individual to manage design, or it would be lost.

At the Libbey Glass Company, Libbey had created a new type of glass craftsman with the title of designer. The designer was the modern reflection of the master craftsman or artist. The designer created a model and then followed it through the manufacturing process. Way back at New England Glass, a young Edward Libbey had offered rewards to employees for pattern designs. He patented designs as art. Libbey's first trip to Europe was spent with engravers and glass artists. While he saw art and manufacturing as related, he saw the role of an art museum in an industrial city as inspirational to its manufacturers. In 1911, Libbey Glass designer Robert Corl defined that relationship:

> To anyone in industrial art, especially in the field of design, a museum is a necessity, and to not take advantage of a large library and the well arranged collections is a very grave mistake for anyone who wishes to thoroughly understand his work, as it contains a record of the best which has been done in the past, and this is an absolute necessity for the production of good work in the future.... The museum holds unlimited ideas for the milliner, the window trimmer, the dressmaker, the arts and crafts worker, the china decorator, the decorator, and in fact everyone who uses the laws of applied art.... In this industrial age success means knowledge, and anyone needing help along the lines of applied art will find material at the Toledo Museum, of which they should not fail to take advantage.[5]

Libbey had helped to create a museum celebrating man as a manufacturer and industrial artist. His glass collection evolved into a historical record of the production of glass, including some of the earliest glass from the fourteenth century B.C. His early pieces from Egypt, Mesopotamia, and Syria illustrate the earliest roots of glassmaking and some early methods lost for centuries. The collection would inspire generations of industrialists and artists. In the 1960s, it would give birth to the glass art movement of Harvey Littleton and Dominick Labino. The roots of that movement can be seen in the words of George Stevens:

> Man, the Maker, however, utilizes two fundamental properties of matter, form and color, the characteristics of which are governed by the same, apparently complex, but

> really simple, laws whereby harmony results in the universe of worlds ... the main stream of our evolution flows swiftly and surely through those natural channels planned by the Creator of things and shaped by the eager hands of man the Maker. To form and color man the Maker early added another element, ornament, with which to embellish his works and ornament responds also to the fundamental law. Art is that science whose laws applied to all things made by man, make them most pleasing to the senses. Art, in its highest expression is disclosed in works of great masters of painting and sculpture but these same laws observed by the masters may be made to apply to all the creations of man the Maker. As they have been utilized in the works of the master they are proof of the great truth, of the mighty yet unharnessed force, which in its application will transcend the benefits of even steam and electricity.[6]

From Libbey's and Stevens's definition of art came the concept of industrial design in the 1930s.

An early trustee, Charles Ashley, saw the same connection between art and manufacturing: "The fine taste and skill shown by children ... is not only a surprise in itself; but gives us an idea of the great commercial value that this training will sometime be for the city in the manufacture of all kinds of things having artistic design."[7] Many city industrialists were excited about this linkage, which the Europeans had often considered the root of manufacturing. Robert Corl, Libbey Glass designer, noted in 1911: "I am sure that the great masses of those who could derive the most practical workable ideas — the ideas which are worth money and which could be adapted to their businesses today, to make for a better and more saleable article — are not using the museum and library to the extent that they should."[8] Other craftsmen in the community noted the improvement of the art museum:

> Bartelle S. Hamilton, a Toledo printer, recognized the museum's worth for members of his trade. The printer had to know the importance of harmony of color, outline, and composition. Hamilton believed the museum's collections rendered printers an opportunity to study valuable information useful to the printing occupation. George S. Mills, an architect, testified about the museum's contributions to his profession. Many young draftsmen became architects without proper education in the fine arts. Mills believed the museum library, in conjunction with the art courses offered at the institution, would enable the "aspiring young draftsmen to become a well rounded architect.'"[9]

The local newspaper echoed the same symbiotic relationship among the museum, community, and industry:

> The Toledo Museum of Art is much more than a museum of art. It is a great education institution of far-reaching influence. It is one of the chief bulwarks of enlightened progress in the Midwest and serves the purposes of industry as well as those of culture. If enabled to continue and expand the same liberal policy in the future that has characterized the past, it will immeasurably increase its practical usefulness and become perhaps the most powerful single institution for general good in Toledo.[10]

Libbey's vision of the role of an art museum was not only new in America but revolutionary. By 1924, museum administrators, industrial designers, managers, educators, and marketers from all over the country were going to the Toledo Museum of Art. In 1927, department stores such as Lamson Brothers and LaSalle Company were sending sales people, floor managers, and managers to design classes. Decades later, in 1962, the museum would lead the Studio Glass Movement of Harvey Littleton and Dominick Labino. The seeds of this movement were first sown by Edward Libbey.

Libbey's legacy includes his concept of art and industrial design. In the early 1880s, Libbey offered design awards to his employees in an effort to merge art and industry. Libbey struggled in his own mind to develop this theme in his collection. The concept of design as a confluence of industry, crafts, art, and manufacture evolved for Libbey as the collection evolved. It was clearly influenced by the collecting and donations of Dr. Gunsaulus, which also tended to be a study in art and manufacturing. By 1918, Libbey's glass collection was truly a manufacturing history of glass, the collection at the museum was thoroughly immersed in manufacture and handcrafts. The collection covered metal, glass, ceramics, pottery, sculpture, painting, cloth, and endless other crafts. When the Toledo Museum of Art launched its School of Design in 1918, it took design to a higher level than Libbey's father-in-law had tried to do in the 1870s with a Toledo Industrial Arts school. The School of Design was located in Florence Libbey's uncle's house (William H. Scott). The Museum School of Design offered free classes in art design, applied design, drawing, pottery, painting, home decoration, and other crafts for Toledo schoolchildren, and had over 1,000 students in its first class. Director George Stevens noted another trend: "More and more these evening classes are being attended by workers and makers. Last night one large group from the Toledo plant came into the class: seven heads of departments and the president of the company himself— all will take the theory of design and harmony that they might better understand how to improve their present and future products. These men will take the same course as is offered to hundreds of our children between the ages of ten and fifteen years."[11]

Another director and trustee, John Wiley, owned Overland Company and was a believer in industrial design. Overland automobiles were known for their design. In 1911, when Henry Ford had one model, Overland was offering eighteen different models. More importantly, Overland incorporated beauty into its models. Wiley, like Libbey, encouraged his managers to visit the museum and to take design courses. Libbey became more convinced that art was the foundation of industry. He established special museum days for local industries. The glass industry, more than any other, was a prime example of art and industry. The symbiotic relationship of art and industry was difficult to fully understand, but years later, in the late 1920s and 1930s, the Libbey legacy helped give birth to the industrial design movement.

Chapter Notes

Chapter 1

1. Warren C. Scoville, *Revolution in Glass-Making* (Cambridge, MA: Harvard University Press, 1948), p. 27.
2. E. William Fairfield, *Fire and Sand: The History of the Libbey-Owens Sheet Glass Company* (Cleveland: Lezius-Hiles, 1960), p. 6.
3. *The Wheeling Daily Intelligencer,* September 14, 1886.
4. MSS 066, University of Toledo Canaday Center Archives, , box 7, folder 28.

Chapter 2

1. The name was originally spelled Libby but an "e" was added in the 1800s.
2. Charles T. Libby, *The Libby Family in America, 1602–1881* (Portland, ME: B. Thurston & Co., 1882), p. 102.
3. Martha Louise Swan, *American Cut and Engraved Glass of the Brilliant Period in Historical Perspective* (Des Moines: Wallace-Homestead Books, 1986), p. 28.
4. Warren C. Scoville, *Revolution in Glass-Making* (Cambridge, MA: Harvard University Press, 1948), p. 57.
5. Deming Jarves, *Reminiscences of Glass-Making* (New York: Hurd & Houghton, 1865), p. 86.

Chapter 3

1. *China, Glass, and Lamps,* July 20, 1905.
2. George and Helen McKearin, *American Glass: The Fine Art of Glassmaking in America* (New York: Crown Publishers, 1941), p. 419.
3. Theodore MacManus, *A Century of Glass Manufacturing, 1818–1918* (Toledo: Libbey Glass, 1918), p. 24.
4. United States Census Office, *Tenth Census of the United States, 1880,* p. 960.
5. Martha Louise Swan, *American Cut and Engraved Glass of the Brilliant Period in Historical Perspective* (Des Moines: Wallace-Homestead Books, 1986), p. 13.
6. Warren C. Scoville, *Revolution in Glass-Making* (Cambridge, MA: Harvard University Press, 1948), p. 31.
7. *The New York Times,* July 17, 1888.
8. Ralph Heller, "Edward Drummond Libbey: Gentleman, Citizen, Pioneer Glassmaker," thesis, University of Toledo, 1948.
9. *The New York Times,* July 17, 1888.

Chapter 4

1. Victor Clark, *History of Manufactures in the United States, 1607–1860* (New York: McGraw-Hill, 1929), p. 517.
2. *Toledo Blade,* July, 25, 1885.

3. *Toledo Blade*, April 17, 1884.
4. Ralph Heller, "Edward Drummond Libbey: Gentleman, Citizen, Pioneer Glassmaker," thesis, University of Toledo, 1948, p. 15.

Chapter 5

1. Not to be confused with his father, Valentine Hicks Ketcham.
2. *Toledo Blade,* September 3, 1887.
3. *Chicago Tribune*, October 2, 1887.
4. Jack Paquette, *Blowpipes: Northwest Ohio Glassmaking in the Gas Boom of the 1880s* (n.p.: Xlibris, 2002), p.337.
5. *Toledo Blade,* March 27, 1888.
6. *Toledo Blade,* February 2, 1889.
7. *Toledo Blade*, March 1, 1890.
8. *Toledo Blade,* May 8, 1888.

Chapter 6

1. Sandra Knudsen, *Libbey Legacy* (Toledo: Toledo Museum Library, 1989), p. 45.
2. *Toledo Blade*, April 17, 1888.
3. *Electrical Engineer* 7 (October 1888).

Chapter 7

1. Warren C. Scoville, *Revolution in Glass-making* (Cambridge, MA: Harvard University Press, 1948), p. 277 and footnote 19.
2. *Toledo Blade,* August 18, 1888.
3. Edward Meigh, "The Development of the Automatic Glass Bottle Machine: A Story of Some Pioneers," *Glass Technology* 1:1 (February 1960). See also E. William Fairfield, *Fire and Sand: The History of the Libbey-Owens Sheet Glass Company* (Cleveland: Lezius-Hiles, 1960), p. 32.
4. Quentin Skrabec, *Michael Owens and the Glass Industry* (Gretna, LA: Pelican, 2007), p. 41.
5. Joseph H. Locke and Jane T. Locke, *Locke Art Glass* (New York: Dover, 1987), p.8.
6. Pearce Davis, *The Development of the American Glass Industry* (Cambridge, MA: Harvard University Press, 1949), p. 160.
7. *Toledo Blade*, June 25, 1890.
8. *The Findlay Republican,* April 29, 1891; see also Jack Paquette, *Blowpipes: Northwest Ohio Glassmaking in the Gas Boom of the 1880s* (n.p.: Xlibris, 2002), p. 97.
9. Solon Richardson deposition, Libbey Glass Company vs. Toledo Glass, November 1, 1920, University of Toledo Canaday Center Archives, MSS 066.

Chapter 8

1. "Applied Arts at the World's Fair," *The Art Amateur* 6 (November 1893), p. 146.
2. "Notes from the World's Columbian Exposition," *Scientific American*, June 24, 1893, p. 387.
3. Kenneth M. Wilson, *American Glass* (New York: Hudson Press, 1994), p. 831.
4. Sandra Knudsen, *Libbey Legacy* (Toledo: Toledo Museum Library, 1989).

Chapter 9

1. Stefan Lorant, *Pittsburgh: The Story of an American City* (Lenox, MA: Donnelley & Sons, 1964), p. 551.
2. John Biggers oral history, University of Toledo Canaday Center Archives, MSS 066, box 7, folder 26.

Chapter 10

1. Toledo Glass log book, , University of Toledo Canady Center Archives, MSS 066, Libbey-Owens-Ford, box 2, folder 1.

2. Jack Paquette, *Blowpipes: Northwest Ohio Glassmaking in the Gas Boom of the 1880s* (n.p.: Xlibris, 2002), p. 350.
3. Solon Richardson deposition, Libbey Glass Company vs. Toledo Glass, November 1, 1920, University of Toledo Canaday Center Archives, MSS 066.
4. Warren C. Scoville, *Revolution in Glass-Making* (Cambridge, MA: Harvard University Press, 1948), p. 327.

Chapter 11

1. Jack Paquette, *Blowpipes: Northwest Ohio Glassmaking in the Gas Boom of the 1880s* (n.p.: Xlibris, 2002), p. 343.
2. "The Genius of the Bottle," *The Toledo Blade*, November 1, 1934.
3. Program, dedication of the Owens AR bottle machine as an international historic engineering landmark by The American Society of Mechanical Engineers, May 17, 1983, online at ASME.org.
4. Ibid.
5. E. William Fairfield, *Fire and Sand: The History of the Libbey-Owens Sheet Glass Company* (Cleveland: Lezius-Hiles, 1960), p. 54.
6. Sandra Knudsen, *Libbey Legacy* (Toledo: Toledo Museum Library, 1989), p. 24.
7. Knudsen, p. 15.
8. Knudsen, p. 59.
9. Knudsen, p. 50.
10. Letter from E. D. Libbey to Nattie Litscher, November 17, 1903, Toledo Museum Archives, box A, shelf 7, file 13.
11. Morgan Barclay, "The Toledo Museum of Art," thesis, University of Toledo, August 1972.

Chapter 12

1. Warren C. Scoville, *Revolution in Glass-Making* (Cambridge, MA: Harvard University Press, 1948), p. 295.
2. *Louisville Courier-Journal,* May 23, 1909.
3. Nina Stevens, *A Man and a Dream: The Book of George W. Stevens* (Hollywood: Hollycrofters, 1941), p. 108.
4. *Museum News,* April, 1907.
5. *The New York Times,* March 17, 1916.
6. Letter from Robinson Locke to Edward Libbey, June 4, 1914, Rutherford Hayes Presidential Library, Robinson Locke Papers, scrapbook.

Chapter 13

1. Nina Stevens, *A Man and a Dream: The Book of George W. Stevens* (Hollywood: Hollycrofters, 1941), p. 13–14.
2. Keene Sumner, "Don't Try to Carry the Whole World on Your Shoulders," *American Magazine,* July 1922.
3. Hugh Hindman, *Child Labor, an American History* (Armonk, NY: M.E. Sharpe, 2002), p. 143.
4. *The New York Times,* September 25, 1907.
5. Labor records, Tiffin Glass Company Collection, Center for Archival Collections, Bowling State University, MS 382, volumes 87–90.
6. Department of Labor, Bureau of Labor Statistics, *Productivity of Labor in the Glass Industry,* bulletin no. 441 (1927).
7. Warren C. Scoville, *Revolution in Glass-Making* (Cambridge, MA: Harvard University Press, 1948), p. 321.
8. Hindman, p. 145.
9. U.S. Bureau of Labor, *Report on condition of woman and child wage-earners in the United States, Volume 3: Glass Industry* (Washington, D.C.: Government Printing Office, 1911), p. 195.
10. *Toledo Museum News* 1:3 (January 1908).
11. *The New York Herald,* February 21, 1909.
12. *The New York Times,* August 29, 1909.

13. *Toledo Museum News* 3:1 (November 1909).
14. *Toledo Times,* March 11, 1911.
15. *The New York Times,* May 18, 1912.
16. Sandra Knudsen, *Libbey Legacy* (Toledo: Toledo Museum Library, 1989), p. 55.
17. Letter from Charles Ashley to Edward Green, July 10, 1908, Toledo Museum of Art Archives.
18. *Toledo Times,* December 25, 1910.

Chapter 14

1. Barbara Floyd, Ann Bowers, and Kimberly Brownlee, "Time in a Bottle: A History of Owens-Illinois Inc., " collection catalog, Ward M. Canaday Center for Special Collections, University of Toledo, 2006.
2. *The New York Times,* January 18, 1912.
3. Sandra Knudsen, *Libbey Legacy* (Toledo: Toledo Museum Library, 1989), p. 11.
4. *Museum News* 19 (March 1912).
5. Ibid.
6. Ibid.
7. Knudsen, p. 56.

Chapter 15

1. Pearce Davis, *The Development of the American Glass Industry,* (New York: Russell & Russell, 1949), p. 127.
2. *Toledo Blade,* September 20, 1906.
3. Warren C. Scoville, *Revolution in Glass-Making* (Cambridge, MA: Harvard University Press, 1948), p. 308.
4. *National Glass Budget* 27 (August 17, 1912).
5. Scoville, p. 288.
6. Ibid.
7. *Museum News* 26 (December 1915).

Chapter 16

1. E. William Fairfield, *Fire and Sand: The History of the Libbey-Owens Sheet Glass Company* (Cleveland: Lezius-Hiles, 1960), p. 34.
2. Warren C. Scoville, *Revolution in Glass-Making* (Cambridge, MA: Harvard University Press, 1948), p. 285.
3. Scoville, p. 304.
4. University of Toledo Canaday Center Archives, MSS 066, box 7, folder 28.

Chapter 17

1. Warren C. Scoville, *Revolution in Glass-Making* (Cambridge, MA: Harvard University Press, 1948), p. 148.
2. *Museum News* 48 (March, 1926).
3. Address by Edward Libbey to the Congress of Art History, Paris, September 26, 1921. University of Toledo Canaday Center Archives, MSS 066, box 8, folder 6.
4. Dennis P. Doorden, *The Alliance of Art and Industry: Toledo Designs for Modern America* (Toledo: Toledo Museum of Art, dist. by Hudson Hills Press, 2002), p. 12.
5. Solon Richardson deposition, Libbey Glass Company vs. Toledo Glass, November 1, 1920, University of Toledo Canaday Center Archives, MSS 066.
6.
7. Ralph Heller, "Edward Drummond Libbey: Gentleman, Citizen, Pioneer Glassmaker," thesis, University of Toledo, 1948, p. 12.
8. Scoville, p. 144.

Chapter 18

1. William S. Walbridge, "The Intimate Lives of Edward Drummond Libbey and Michael Owens as We Knew Them," University of Toledo Canaday Center Archives, Libbey Glass files.
2. Letter from Henry Sprang, April 24, 1964, Toledo Museum Archives, shelf 7, box A, file 11.

Chapter 19

1. William S. Walbridge, "The Intimate Lives of Edward D. Libbey and Michael J. Owens as We Knew Them," December 1, 1933, University of Toledo Canaday Center Archives, Libbey Glass files.
2. Warren C. Scoville, *Revolution in Glass-Making* (Cambridge, MA: Harvard University Press, 1948), p. 307.
3. "Builders of Business," *System,* June, 1919. University of Toledo Canaday Center Archives, MSS 097, box 2, folder 4.
4. Recollections of John D. Biggers, 1967/1968, University of Toledo Canaday Center Archives, Libbey Glass files.

Chapter 20

1. Tariff Commission of Congress, *Report of 1882* (Washington, DC: Government Printing Office, 1882), p. 2153–2513.
2. Warren C. Scoville, *Revolution in Glass-Making* (Cambridge, MA: Harvard University Press, 1948), p. 318.
3. Pearce Davis, *The Development of the American Glass Industry,* (New York: Russell & Russell, 1949).
4. Sandra Knudsen, *Libbey Legacy* (Toledo: Toledo Museum Library, 1989), p. 13.
5. *Toledo Times,* March 14, 1905.
6. *Toledo Times,* January 24, 1909.
7. Knudsen, p. 61.
8. Morgan Barclay, "The Toledo Museum of Art," thesis, University of Toledo, 1972, p. 53.

Chapter 21

1. Regina Lee Blaszczyk, *Imaging Consumers: Design and Innovation from Wedgwood to Corning,* (Baltimore: The John Hopkins University Press, 2000), p. 44.
2. Blaszczyk, p. 45.
3. *Toledo Museum News* 1:1 (November, 1907).
4. Dennis P. Doorden, *The Alliance of Art and Industry: Toledo Designs for Modern America* (Toledo: Toledo Museum of Art, dist. by Hudson Hills Press, 2002), p. 7.
5. Robert Corl, *Toledo Times,* February 17, 1911.
6. *The Toledo Museum of Art News* 38 (February, 1921).
7. Charles Ashley, letter to editor, *Toledo Blade,* May 1, 1906.
8. *Toledo Times,* February 17, 1911.
9. Morgan Barclay, "The Toledo Museum of Art," thesis, University of Toledo, 1972, p. 57.
10. *Toledo Times,* December 25, 1910.
11. *Toledo Museum of Art News* 38 (February, 1921).

Bibliography

Print Sources

Aiken, E. William. *The Roots Grow Deep: A Story of Captain Ford, His Son Edward and Their Contribution to America's Glass Industry, and a Picture of People and Events That Helped to Build Our Land.* Cleveland: Lezius-Hiles, 1957.

American Society of Mechanical Engineers. "International Landmark Award." Toledo, 1983.

Barclay, Morgan. "The Toledo Museum of Art." Thesis, University of Toledo, August 1972

Blaszczyk, Regina Lee. *Imaging Consumers: Design and Innovation from Wedgwood to Corning.* Baltimore: Johns Hopkins University Press, 2000.

Boggess, Bill, and Louise Boggess. *American Brilliant Cut Glass.* New York: Crown, 1977.

Broderick, John. *Forty Years with General Electric.* Albany: Fort Orange Press, 1929.

Clark, Victor. *History of Manufactures in the United States, 1607–1860.* New York: McGraw-Hill, 1929.

Diamond, Freda. *The Story of Glass, from the Pages of* Antiques. New York: Harcourt, Brace, 1953.

Downes, Randolph. *Lucas County Historical Series, Volumes I–IV.* Toledo: Historical Society of Northwestern Ohio, 1951.

Doyle, John. *A Story of Early Toledo.* Bowling Green, OH: C. S. Tassel, 1919.

Fairfield, E. William. *Fire and Sand: The History of the Libbey-Owens Sheet Glass Company.* Cleveland: Lezius-Hiles, 1960.

Fauster, Carl. *Libbey Glass Since 1818: Pictorial History & Collector's Guide.* Toledo: Len Beach, 1979.

Floyd, Barbara, Ann Bowers, and Kimberly Brownlee. "Time in a Bottle." Catalog by Ward M. Canaday Center for Special Collections, University of Toledo, 2006.

Fowle, Arthur. *Flat Glass.* Toledo: Libbey-Owens Sheet Glass, 1924.

Hallenbeck, Thomas. Papers. Lucas County Library, History Room, Toledo, Ohio.

Harrington, J.C. *Glassmaking at Jamestown: America's First Industry.* Richmond: Deitz, 1952.

Heller, Ralph. "Edward Drummond Libbey: Gentleman, Citizen, Pioneer Glassmaker," Thesis. University of Toledo, 1948.

Hodges, Elizabeth. *Manufacture of Glass.* New York: Sterling, 1948.

Husfloen, Kyle. *American Pressed Glass.* Des Moines: Wallace-Homestead, 1992.

Jarves, Deming. *Reminiscences of Glass-Making.* New York: Hard & Houghton, 1965.

Kassel, Karl. "The Glass Industry." Thesis, University of Toledo, 1966.

Knudsen, Sandra. *Libbey Legacy.* Toledo Museum Library, 1989.

Lezius, W. "The Glass Industry in Toledo." Thesis, University of Toledo, 1937.

MacManus, Theodore. *A Century of Glass Manufacture, 1818–1918.* Toledo: Libbey Glass, 1918.

McKearin, George, and Helen McKearin. *American Glass: The Fine Art of Glassmaking I America.* New York: Crown, 1941.

_____, and _____. *Two Hundred Years of American Blown Glass.* New York: Crown, 1958.

McKearin, Helen, and Kenneth Wilson. *American Bottles and Flasks and Their Ancestry.* New York: Crown Publishers, 1978.

Measall, James, Don Smith, et al. *Findlay Glass: The Glass Tableware Manufacturers, 1886–1902*. Marietta, OH: Antique Publications, 1986.
Morgan, H. Wayne. *William McKinley and His America*. Syracuse: Syracuse University Press, 1962.
Paquette, Jack. *Blowpipes: Northwest Ohio Glassmaking in the Gas Boom of the 1880s*. Toledo: Xlibris, 2002.
_____. *The Glassmakers: A History of Owens-Illinois Incorporated*. Toledo: Trumpeting Angel Press, 1994.
Peterson, Arthur. *Glass Patents and Patterns*. Sanford: Celery City Printing, 1973.
Phillips, C. *Glass: The Miracle Maker*. London: Pitman & Sons, 1948.
Porter, Tana. *Toledo Profile*. Toledo: Toledo Sesquicentennial Commission, 1987
Revi, A. C. *American Pressed Glass and Figure Bottles*. New York: Thomas Nelson & Sons, 1973.
Schwartz, Marvin D., and Robert DiBartolomeo. *American Glass*. Princeton: The Pyne Press, 1922.
Scoville, Warren. "Growth of the American Glass Industry to 1880." *The Journal of Political Economy*, vol. LII, no. 3, September 1944.
_____. *Revolution in Glass-Making*. Cambridge, MA: Harvard University Press, 1948.
Skrabec, Quentin. *George Westinghouse: Gentle Genius*. New York: Algora, 2007.
_____. *Glass in Northwest Ohio*. Charleston: Arcadia, 2007.
_____. *The Metallurgic Age: The Victorian Flowering of Invention and Industrial Science*. Jefferson: McFarland, 2006.
_____. *Michael Owens and the Glass Industry*. Gretna, LA: Pelican, 2007.
_____. *William McKinley: Apostle of Protectionism*. New York: Algora, 2008.
Slavishak, Edward. *Bodies of Work: Civic Display and Labor in Industrial*. Durham: Duke University Press, 2008.
Smith, Page. *Rise of Industrial America: A People's History of the Post-Reconstruction Era*. New York: McGraw-Hill, 1984.
Spillman, Jane Shadel. *Glass from World's Fairs, 1851–1904*. Corning, NY: Corning Museum, 1986.
_____. *Glassmaking: America's First Industry*. Corning, NY: Corning Museum, 1976.
Stevens, Nina. *A Man and a Dream: The Book of George W. Stevens*. Hollywood: Hollycrofters, 1941.
Swan, Martha Louise. *American Cut and Engraved Glass of the Brilliant Period in Historical Perspective*. Wallace-Homestead, 1986.
Walbridge, William. *American Bottles*. Toledo: Caslon, 1920.
Waugh, Sidney. *The Art of Glass Making*. New York: Dodd, Mead, 1939.
Welker, John, and Elizabeth Welker. *Pressed Glass in America: Encyclopedia of the First Hundred Years, 1825–1925*. Ivyland, PA: Antique Acres Press, 1985.
Wilson, Kenneth. *American Glass*. New York: Hudson Press, 1994.
_____. *New England Glass and Glassmaking*. New York: Thomas Crowell, 1972.

Archival Sources

Bowling Green State University, Special Collections.
Fostoria Kaubisch Memorial Public Library.
The Henry Ford, Benson Center.
Lucas County Library, Thomas Hallenbeck Papers and Clipping Files.
The Toledo Museum of Art, Library Papers and Museum Records.
University of Toledo, Ward M. Canaday Center: Libbey Glass Records, Libbey-Owen Records and Jack Paquette Papers.

Index